SUCCESS WITH CICHLIDS FROM LAKES MALAWI AND TANGANYIKA

Sabine Melke

Neolamprologus tretocephalus in their natural habitat.

D1400522

Translated by:

U.Erich Friese
General Curator
Sydney Aquarium
Sydney, Australia

All photos by Bernd Melke except for photos showing products and the following: Dr. Herbert R. Axelrod, pp. 54, 59, 60, 61 bottom, 62, 71; Glen S. Axelrod, p. 67 bottom; Pierre Brichard, pp. 63 top, 65, 66, 67 top, 70; Bernd Degen, pp. 9, 28, 33; Dr. Harry Grier, pp. 48, 49; Bernd Kilian, pp. 6, 7, 41, 64, 68, 69, 72, 74, 76; Ad Konings, title page, 85, 179; H.-J. Richter, pp. 93, 147, 175, 176; M. P. & C. Piednoir, pp. 91, 92; C. O. Masters, p. 148; Isabelle Francais, pp. 149, 150, 152; A. Kapralski, pp. 170, 171, 172, 173; W. Ross, p. 178; Heinrich Stolz, p. 63 bottom; Dieter Untergasser, p. 186.

Distributed in the UNITED STATES to the Pet Trade by T.F.H. Publications, Inc., One T.F.H. Plaza, Neptune City, NJ 07753; distributed in the UNITED STATES to the Bookstore and Library Trade by National Book Network, Inc. 4720 Boston Way, Lanham MD 20706; in CANADA to the Pet Trade by H & L Pet Supplies Inc., 27 Kingston Crescent, Kitchener, Ontario N2B 2T6; Rolf C. Hagen Ltd., 3225 Sartelon Street, Montreal 382 Quebec; in CANADA to the Book Trade by Macmillan of Canada (A Division of Canada Publishing Corporation), 164 Commander Boulevard, Agincourt, Ontario M1S 3C7; in the United Kingdom by T.F.H. Publications, PO Box 15, Waterlooville PO7 6BQ; in AUSTRALIA AND THE SOUTH PACIFIC by T.F.H. (Australia), Pty. Ltd., Box 149, Brookvale 2100 N.S.W., Australia; in NEW ZEALAND by Brooklands Aquarium Ltd. 5 McGiven Drive, New Plymouth, RD1 New Zealand; in Japan by T.F.H. Publications, Japan—Jiro Tsuda, 10-12-3 Ohjidai, Sakura, Chiba 285, Japan; in SOUTH AFRICA by Multipet Pty. Ltd., P.O. Box 35347, Northway, 4065, South Africa. Published by T.F.H. Publications, Inc.
Manufactured in the United States of America by T.F.H. Publications, Inc.

In the world of African cichlids and cichlids as a whole, T.F.H. Publications, Inc. has led the way in the production of a multitude of good books. These are by far the most comprehensive and colorfully illustrated books on the subject of cichlids available. They are available at pet shops and tropical fish specialty stores the world over.

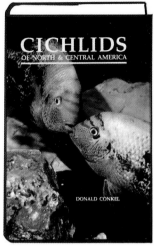
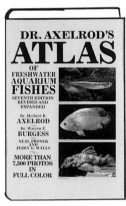

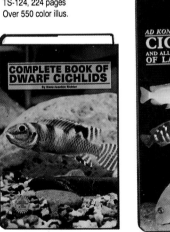
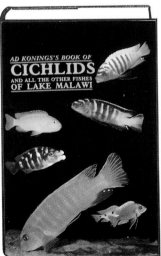
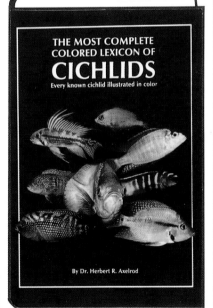

ENGLISH/METRIC CONVERSION CHART

CUSTOMARY U.S. MEASURES AND EQUIVALENTS

METRIC MEASURES AND EQUIVALENTS

LENGTH

1 inch (in)		= 2.54 cm
1 foot (ft)	= 12 in	= .3048 m
1 yard (yd)	= 3 ft	= .9144 m
1 mile (mi)	= 1760 yd	= 1.6093 km
1 nautical mile	= 1.152 mi	= 1.853 km

1 millimeter (mm)		= .0394 in
1 centimeter (cm)	= 10 mm	= .3937 in
1 meter (m)	= 1000 mm	= 1.0936 yd
1 kilometer (km)	= 1000 m	= .6214 mi

AREA

1 square inch (in²)		= 6.4516 cm²
1 sq. foot (ft²)	= 144 in²	=.093 m²
1 sq. yard (yd²)	= 9 ft²	= .8361 m²
1 acre	= 4840 yd²	= 4046.86 m²
1 sq. mile(mi²)	= 640 acre	= 2.59 km²

1 sq centimeter (cm²)	= 100 mm²	= .155 in²
1 sq meter (m²)	= 10,000 cm²	= 1.196 yd²
1 hectare (ha)	= 10,000 m²	= 2.4711 acres
1 sq kilometer (km²)	= 100 ha	= .3861 mi²

WEIGHT

1 ounce (oz)	= 437.5 grains	= 28.35 g
1 pound (lb)	= 16 oz	= .4536 kg
1 short ton	= 2000 lb	= .9072 t
1 long ton	= 2240 lb	= 1.0161 t

1 milligram (mg)		= .0154 grain
1 gram (g)	= 1000 mg	= .0353 oz
1 kilogram (kg)	= 1000 g	= 2.2046 lb
1 tonne (t)	= 1000 kg	= 1.1023 short tons
1 tonne		= .9842 long ton

VOLUME

1 cubic inch (in³)		= 16.387 cm³
1 cubic foot (ft³)	= 1728 in³	= .028 m³
1 cubic yard (yd³)	= 27 ft³	= .7646 m³
1 fluid ounce (fl oz)		= 2.957 cl
1 liquid pint (pt)	= 16 fl oz	= .4732 l
1 liquid quart (qt)	= 2 pt	= .946 l
1 gallon (gal)	= 4 qt	= 3.7853 l
1 dry pint		= .5506 l
1 bushel (bu)	= 64 dry pt	= 35.2381 l

1 cubic centimeter (cm³)		= .061 in³
1 cubic decimeter (dm³)	= 1000 cm³	= .353 ft³
1 cubic meter (m³)	= 1000 dm³	= 1.3079 yd³
1 liter (l)	= 1 dm³	= .2642 gal
1 hectoliter (hl)	= 100 l	= 2.8378 bu

TEMPERATURE

CELSIUS° = 5/9 (F° − 32°)   FAHRENHEIT° = 9/5 C° + 32°

CONTENTS

Lake Malawi habitat tank. Sedimentary rocks were used.

PREFACE

ABOUT THE USE OF THE GENERIC NAME *HAPLOCHROMIS*

In the latest revision by D. Eccles and Dr. E. Trewavas, the Malawi *Haplochromis* species have been reclassified into 23 new and 3 previously

established genera. Since this revision is considered to be preliminary, the name *Haplochromis* has been placed in quotation marks in this book. Moreover, using this more familiar generic name makes it easier for cichlid fanciers to search the relevant literature for information about particular species.

Cichlids endemic to the East African lakes have already many devotees among home aquarists and the number of enthusiasts continues to grow steadily. There are no other cichlids that are so popular as those from Lakes Malawi and Tanganyika. These—in terms of their evolutionary development—highly advanced fishes have stimulated aquarists' interest with their broad behavioral spectrum and fantastic shapes and colors. This book has been written for beginners as well as for experienced aquarists who want to broaden their knowledge about these cichlids. Within that context it is a reference text that answers many questions about cichlid care and maintenance. Providing habitat- and species-correct care is the basic aim of this book. At the same time, it is meant to enable you to fully enjoy these incredible fishes, a joy that I have been able to participate in for many years.

At this stage I would like to express my thanks to Mr. Bernd Kilian, who contributed important habitat photographs to this book. Similarly, the same appreciation is being extended to Mr. Bernd Degen. Without him this book would not have been written. He was always available for consultation and advice. I would also like to convey my thanks to my husband, who helped me, with his photographs, to realize this book in just the way I wanted it to be.

I do hope that you will enjoy this book and that it becomes a valuable aid for you when keeping these cichlids.

Sabine Melke

THE AQUARIUM

CORRECT LOCATION

Several points have to be considered when acquiring and setting up a fish tank. First of all, the location is of paramount importance. This should be a quiet, shaded spot. Cichlids can be very sensitive to an unsettled home environment, which could lead to a so-called "panic" aquarium. Totally unsuited for an aquarium are those locations where people frequently and suddenly appear, such as hallways, near doorways, and (if there are small children around) in nurseries. If such an unsettled location cannot be avoided the tank should be screened off.

This can be accomplished in a number of ways. First, the sides of the tank can be covered with special decorative aquarium foil. Also, potted plants placed strategically along both sides of the tank can provide an adequate sight barrier to protect the fishes. We also have to avoid a sunny location for the tank, because this would enhance undesirable algal growth and overheat the water. The current trend is to set up tropical fish tanks among extensive stands of potted plants in enclosed verandas or so-called winter gardens. This is indeed quite possible provided the above-mentioned guidelines are taken into consideration. Yet, having to provide adequate supplementary aeration cannot be avoided, especially in locations that are exposed to lots of sun. If in doubt, it is advisable to check and record the maximum temperatures in the proposed location on hot days. Locations that have air temperatures clearly in excess of 30°C during the summer are unsuitable for fish tanks.

You must also pay attention to the floor. A concrete floor is more suited for the placement of a fish tank than a less solid wooden floor. Footsteps trigger vibrations in wooden floors that are then transmitted to the aquarium. Fishes can "feel" these vibrations since they are transmitted through water. This can start a panic. This effect can be mitigated somewhat by placing a soft material (e.g. styrofoam layer) between the tank and the stand it is placed on. Moreover, tanks on wooden floors should be

Facing page: The sides and back wall of this Lake Malawi tank have been painted blue on the outside.

A tank should be placed on a shock-absorbing sheet of resilient material. This cushions any mechanical stress and tension once the tank has been filled.

placed in such a way that the entire weight (distributed over the legs or other support of the stand) is positioned directly over the major timbers supporting the floor boards. This lessens the effect of vibrations and provides better support (i.e. better load distribution).

WHAT SIZE TANK?

In order to get maximum enjoyment from the care of East African cichlids you should select a tank that is as large as possible. The ideal size is a tank of not less than 1.5 m length. This size tank enables you to keep nearly any cichlid species you desire. Smaller size tanks can, of course, also be used, although this limits the number of species that can be kept. In essence, then, the smaller the tank, the smaller the selection of suitable species that can be kept in it. For example, only a few snail cichlids (*Neolamprologus* spp.) and slender cichlids (*Julidochromis* and *Chalinochromis* spp.) are suitable for small tanks. But even these species should not be given tanks smaller than about 50 cm in length.

THE AQUARIUM STAND

Cichlid enthusiasts tend to use mainly medium to large tanks. So that it can support the massive weight of such an aquarium full of water, the aquarium stand must be correspondingly strong. The larger the tank the more important it becomes that the stand is of the proper type of construction and sufficiently solidly built. Simple aquarium stands as well as enclosed cabinet stands are commercially available in many sizes, colors, and designs. These professionally built, brand-name stands are designed to carry the weight of a fully set-up aquarium. Any aquarist with special requirements with regard to tank stand or aquarium cabinet can, of course, build it himself or contract a tradesman. Special sizes of tanks and stands are the "foster children" of commercial manufacturers.

A relatively simple alternative as an aquarium stand—even for those not skilled with their hands—can be the following design based on lightweight brick construction. These bricks are characterized by reduced weight as well as high material strength.

A painted sheet of plastic foam has been used a a backdrop. This enhances the appearance of the fishes. At the lower left is *Cryptocoryne usteriana* (a large female plant).

Moreover, the shape and size of such dressed, lightweight bricks can easily be modified with a hand saw. These bricks are available in a number of sizes to meet all requirements. First you determine the actual height you want for your tank. Keep in mind, though, that you should be able to watch your fishes at eye level from a sitting position. The minimum height for such an aquarium stand should be about 70 cm. Using lightweight bricks you can set up columns as a tank support substructure. The actual length of the bricks to be used depends on the width of the aquarium. The width of the bricks should be at least 20 cm, and the height of the bricks can be variable, depending upon the final height to be achieved. Tanks with a length of up to 1.2 m require two columns, at 1.5 m we need three columns, and for tanks from about 1.8 m length up there should be four brick columns in place.

If necessary, the bricks can be held together with a special cement that is also available from building material suppliers, but this has no influence on the stability of the columns. It is recommended, however, that the columns be tiled for esthetic purposes. If you are not going to tile the column you should install some means of closing off the space between adjacent columns and then use it as storage space for filters, electric wires, and similar items. Wooden panels, plywood, masonite, and similar materials can be used. Most building material suppliers have a wide range of decorative panels available that will blend in with any modern living style. Also useful are wooden veneers as well as natural wallpapers (with wood, cork, and rock motifs) and adhesive foils. You can give your imagination a free run.

The "doors" are attached by means of door magnets (available from most hardware stores that carry building materials). Two magnets are placed on the inside of the doors and the other two matching magnets are placed on the stand or column structure. They are attached on the stand in exactly the position where the door closes off the vacant space under the tank. This then disguises all visible sides of the aquarium substructure. The door magnets facilitate quick access to the space below the aquarium.

The stand (more

appropriately described as the aquarium substructure) described here is particularly suited for wooden floors, since it minimizes the transmission of floor vibrations, facilitated by the wide support surface area of these bricks. At the same time the total weight is spread over a larger floor area, in contrast to the high point load created by traditional aquarium stands with individual legs. Apart from its many other advantages, this aquarium support structure is also very cost-effective, particularly for large aquariums.

POSITIONING THE TANK

Large tanks with a volume of 350 liters or more are best secured to a load-bearing wall, which is characterized by greater wall thickness than intermediate walls. As a rule, outer walls are usually load-bearing walls. In addition, the load-carrying capacity varies among individual floors and intermediate (false) ceilings. If in doubt, you should seek professional advice. This, however, applies only to large tanks (from 350 liters up) and if it involves an older building. For yet larger tanks you must seek information about the load/carrying capacity of the floor/ceiling. Generally, however, one would not expect any difficulties with such a load. The load-carrying capacity need not be taken into consideration when setting up a large tank in a basement, since there will be adequate structural support from the concrete building foundation.

Placement of the tank must be done with care and deliberation. This is of paramount importance if

A cichlid-lover's dream is this 1000 gallon double bull nose aquarium with steel stand and complete cabinetry, made by American Acrylic, here shown housing marine fishes.

unpleasant surprises are to be avoided at a later date. A spirit level is essential for positioning a tank. Uneven surfaces are quickly recognized and the necessary adjustment can easily be made. Ready-made aquarium cabinets usually come with adjustable legs that can compensate for uneven floor areas. If, however, you have opted for aquarium support columns assembled from bricks, uneven floor areas will have to be leveled out with styrofoam, paper, cardboard, or similar materials. These aids can be used on the floor underneath the bricks or as the last layer on top of the bricks.

Finally, we place a wooden board of about 20 mm thickness or more on top of the columns. You can select a solid finished board or use plywood or even a sheet of composite marerial (however, the latter tends to swell up when exposed to moisture). The wooden board serves to provide a level plane as support for the aquarium. Once the board has been put in place it must be checked again with the spirit level. Readings must be taken not only along the width and length but also across the diagonals. If

need be, corrections and/or modifications must be made until all sides conform. Once the spirit level gives the "green light" a thin sheet of styrofoam or foam rubber is placed over the board. This material will compensate for the slightest uneven spots, such as sand grains or protruding silicon seals.

Once this preliminary work has been completed the real task of setting up the tank can begin.

BUYING THE TANK

First a few words about commercially available tanks in general. You will notice that there are various glass qualities, gluing techniques, and designs. Fundamentally, we distinguish between all-glass tanks and the outmoded type in which individual glass panels have been secured inside a metal frame. The former are clearly preferred, since that sort of design permits all-around visibility into the tank without being impeded by the tank frame.

In small to medium-size tanks the individual sheets of glass are glued (with silicone) at their outer edges only. Larger tanks have often an added silicone seam placed in each corner for added safety. The advantages and disadvantages of both techniques are debatable. I personally prefer the former method, which provides for a neater seal. The latter method uses the softer

A group of *Tropheus moorii* 'Ikola'. None of the individuals are being suppressed when a group is sufficiently large. All fishes display their most striking coloration.

This double bubble aquarium was manufactured and designed by the American Acrylic Manufacturing Company of San Diego, California. It is about 125 x 60 x 250 cm (48 x 22 x 96 inches) and comes complete with all the necessary filters, aerators, etc.

silicone, and the interface between glue and glass is often invaded by algae. This reduces the longevity as well as compromises the safety of the tank. I was able to observe this process once over a period of a few years, although the tanks were not exposed to solar radiation. Even using differently colored silicones did not change the effect. The further algal penetration advances, the more the stability of the glued surfaces becomes impeded. For that reason I recommend to you the use of tanks that have been firmly glued with a solid silicone. This material cannot be undermined by algae.

When selecting a tank look for meticulous workmanship. The silicone adhesive areas must be uniform and without air pockets. Tanks in excess of 1.2 m length should have a center brace for added stability. Information about glass quality is available from the trade. In order to avoid unpleasant distortions, you should select tanks made out of crystal (plate) glass. You do not have to be concerned about particular glass thicknesses as long as you purchase your tank from a reputable dealer. Glass thickness required for a particular size tank will have been taken into consideration during manufacture. Some tank sizes, however, are available in two different glass thicknesses. Of course, the thicker glass is also more expensive, but it also affords greater safety.

When buying a second-hand tank, the above-mentioned points must also be taken into consideration. A previously owned tank may be momentarily a cheaper acquisition, but its longevity is also reduced. Therefore, used tanks should be checked for leaks and you must find out how old such a tank really is. After having been in use for 8 to 10 years, a tank should be thoroughly checked. This should include close scrutiny of the center brace. All glued surfaces should be examined for leaks.

Well-made castles are decorative and useful. The cichlids hide inside the castles and feel very secure there. These castles are available at many petshops and are made by Blue Ribbon.

BLUE RIBBON PET PRODUCTS, INC.

A small tank like this (50 liters) with snail cichlids such as *Neolamprologus ocellatus* has a certain charm.

Deficient areas must be repaired with a special aquarium silicone cement.

Standard size tanks are generally cheaper than tanks with special dimensions. Unfortunately, tanks with a depth (front to back) of 60 cm are generally not part of the range of standard tanks available. Yet, tanks with such a depth cater optimally to the territorial demands of cichlids. These fishes require a greater area than a particular water height. An additional 10 cm depth provides more space for territories and affords a greater opportunity for esthetically pleasing landscape work. Therefore, you should give preference to a tank with maximum (affordable) width.

Before actually using the tank it should be "leached" for 2 to 3 days. This tends to dissolve silicone remnants. At the same time, this is an opportunity to check the tank for any leaks.

TANK COVER

An aquarium must have a cover, either as part of the aquarium cabinet structure with a built-in source of illumination or just simply a sheet of glass. There are a number of

variations available from aquarium shops. The tops of aquarium cabinets invariably have lights integrated in them. This sort of arrangement is extremely practical and makes servicing the tank very easy. The light weight facilitates a quick removal of the entire tank cover. Beyond that, there are models where lids and lights are removable via sliding tracks. This also makes servicing an aquarium much more convienient.

Another method utilizes suspended lamps operated by means of a retractable cord and pulley arrangement. A single manual action elevates the lights sufficiently high above the tank to provide unimpeded access. In this case, the top of the tank should be covered with a sheet of glass to avoid having fish jump out.

The tank can also be closed off with an open wooden frame. This variation is particularly useful if potted plants are used for external decoration.

You can, of course, construct a wooden box cover yourself, but I should point out that this does require some carpentry skills. Wooden boards may be pre-cut to your specifications and assembled as per your requirements. A hinged lid (door hinge!) permits service access to the tank and lights. A narrow strip of timber—as a retaining ledge—around the inside of the entire box provides a firm seat around the top of the tank. This should be fastened in a such a way that it covers the water level once the frame is in position on top of the tank.

Fluorescent tubes are then installed inside the tank top to provide lighting. Make sure to use only light fittings approved for a wet environment (outdoor fittings!). Again, the actual tank should be covered with a sheet of glass (or several smaller sheets for easier access). I must, however, be honest and admit that home-made tank tops are invariably quite heavy, which makes them difficult, or at least awkward, to handle. Therefore, you are better advised to give preference to commercially available aquarium covers.

LIGHTING

Lighting for a cichlid tank does not need to be excessive. In fact, most cichlids react unfavorably to too much light. Moreover, East African cichlid tanks supposed to

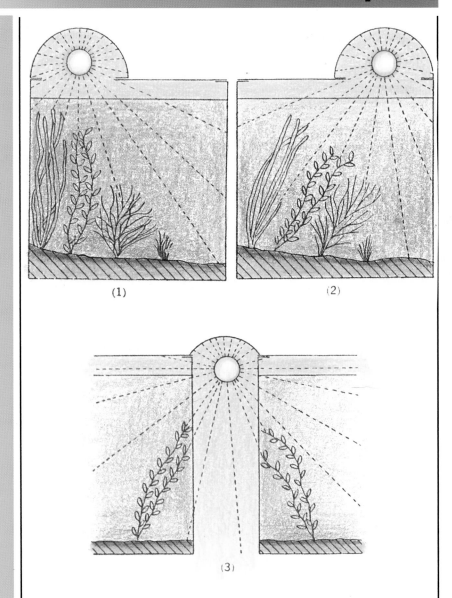

(1)

(2)

(3)

PROPER LIGHTING

(1) Plants are positively phototropic. This means they bend toward the light source. This is the proper lighting scheme, with the light source situated over the plants.

(2) When the light source is placed in front (to facilitate viewing the fishes), the plants will bend toward the light source as shown here.

(3) Some hobbyists have a rack of tanks, one next to the other, and they decorate the tanks with heavy plantings on the ends of the tanks near the light source.

be environmentally correct will have only a moderate number of plants. Therefore, only two fluorescent tubes are all that will ever be required. In fact, tanks up to 80 cm in length will really only need a single fluorescent light.

Selecting the actual type of light is very much a matter of individual esthetic taste. Yet if you do not want to distort the true colors of the cichlids, you should use daylight fluorescent tubes. This sort of light is bright and whitish and gives a natural appearance to the colors of the tank inhabitants. Special plant lights need not be used. These not only enhance the growth of desired plants but also that of undesirable ones. Algae will grow particularly well under plant lights because they are plants too. So you are best advised to do without special plant lights if you do not want algae in your tank. The other plants will grow anyway, but maybe not as fast.

HEATING

Now let us turn to some of the technical aspects. First of all, we have to make provisions for heating when keeping tropical freshwater fishes. Very practical and highly reliable for everyday use are the thermostatically controlled rod heaters. These are available from a number of manufacturers. When buying such a heater make sure that the end cap is properly fitted to protect the heater's interior by preventing water from entering. Here it is important to pay more attention to quality than to the actual price. Some automatic heaters are available with a temperature adjustment and an electric cut-out device to prevent

Top, left: The pump should be located in a position higher than the tank. If the pump breaks and the water siphons out, the entire tank could be emptied if the pump is lower than the tank. **Top, right:** Undergravel filters often are used in conjuunction with disposable charcoal filters that attach to the return stem. **Bottom, left:** Petshops offer full hood reflectors; they cover the aquarium completely and contain the necessary lighting. **Bottom, right:** The full hood has a fixed position for the light, so you must decorate your aquarium accordingly or the plants will grow in an unsatisfactory manner.

overheating when a plugged-in heater is not emersed in water. This facilitates easy handling and a correct temperature setting. At the same time

liters of tank water.

Another alternative is filters with a built-in (integrated) heater. As practical as these may be, there are still

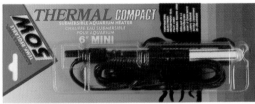

the cut-out device compensates for any carelessness.

Always select an automatic heater of the correct wattage to heat a particular sized tank. The technical details for this come with the operating instructions for these heaters. A heater that is too large for a particular tank can easily cook your fishes if it becomes defective. But this can, of course, only happen if the heater that has been selected is too powerful to start with. Therefore, always pay close attention to the correct heating capacity. If the manufacturer's instructions are not available, calculate the required heating capacity as follows: 1 watt per 2

disadvantages: If one of the two functions (heater or filter) fails, the entire unit has to be replaced or sent in for repairs.

I must advise against the use of heating mats in cichlid tanks, especially when styrofoam is to be placed on the interior tank floor. Under these circumstances a heat blockage could develop and the increased heat could crack the glass. In addition, defective heating mats are difficult to exchange, because usually the entire tank will have to be dismantled.

If automatic rod heaters are used, one can also install two of them for safety in the event one of them breaks down. Irrespective of what heating

If there is one piece of equipment that should be of especially high quality it is the *heater.* A broken heater can cook the fishes or allow them to freeze. It might even explode if water seeps in and turns to steam when it hits the heater coil. The two Hagen heaters shown here are of the submersible type.

method is being used, it is always important to monitor the temperature on a regular basis. This is necessary to prevent incorrect temperatures from causing serious damage and injuries to the fishes. A clearly visible thermometer stuck to one side of the front or side viewing panels will convey at a single glance the current water temperature of the tank.

WHAT TYPE OF FILTRATION?

The centerpiece of technology for an aquarium is its filtration. Since East African cichlids occur in ocean-like currents, we have to take this into account. Therefore, sufficient water turnover contributes significantly to the well-being of our charges. Consequently, we need a powerful pump for circulating large volumes of water. A number of different circulating pumps have been brought onto the market by various manufacturers. When operated in conjunction with a filter (pressure) vessel or a filter cartridge they produce a current as well as pre-filtration of the water. Water turnover (circulation) can be increased several times over, which further improves the well-being of your fishes. A strong current assures vitality and appetite and enhances immune responses, which, in turn, promotes growth. Any observer will find greater pleasure in viewing such agile, active fishes.

To supplement mechanical filtration (as described above) it is recommended that you also install biological filtration, which actively removes metabolic waste products dissolved in the aquarium water. Both types of filtration (mechanical and biological) have a role to play. Aerating filters are particularly well suited for that purpose. Unfortunately, there are only a few aerating filters of this type available. They have proven to be highly effective in aquarium operations but, unfortunately, aquarium shops rarely ever have these filters for sale. Instead, this sort of system has most of its devotees among breeders. Aerating filters remove tiny dirt particles from the water

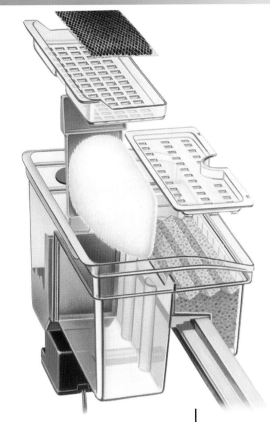

and enrich the water with dissolved oxygen by means of very fine bubbles. Large air bubbles, as produced by many other filters, do not contribute significantly to an oxygen-enrichment of the water. Instead, large air bubbles flow too rapidly to the tank surface. Water sufficiently enriched with oxygen is necessary for the activities of aerobic bacteria, which facilitate the reduction of metabolic waste products. Aerating Filters therefore cater to the needs of East African cichlids, which require sufficient water flow, and also oxygen-rich, crystal

This Willinger power filter increases its effectiveness by containing a number of different filtration media that trap contaminants in a wide range of particle sizes.

Submersible power filter heads, like the Rolf C. Hagen models shown here, will greatly enhance the filtering capacity of biological undergravel filters.

clear water for their well-being.

Another alternative for biological filtration is in the integration of a biological filter chamber. For that purpose a bio-chamber may be glued into the aquarium. This consists of three separate sub-chambers that are divided by glass partitions without inhibiting water flow. The first chamber serves as a receptacle to accept siphoned water from the tank, which transports sediments into this chamber. The large sedimentary components settle out on the bottom of this chamber. From there they can be siphoned out manually during the next water change, before they reach the filter medium. The next (second) chamber is filled with filter medium, which can be various inert plastic or ceramic bodies and filter wool. This chamber can also be completely filled with wool. In any event, the top layer should be filter wool, which retains the solid dirt

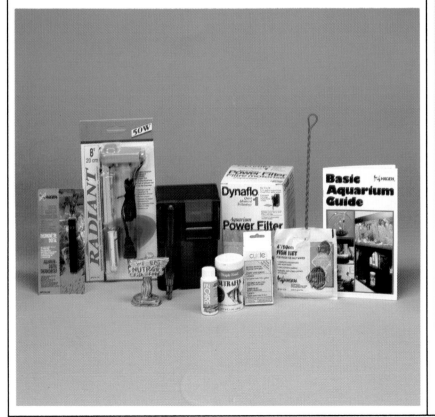

The two Hagen starter kits shown here are examples of the money-saving complete aquarium setups (note that the kit shown in the upper photo contains a tank as well as the equipment) available at pet shops and tropical fish specialty stores.

These efficient filters and pumps are made by Willinger Bio® and are available at most petshops.

particles and mechanically cleans the water. The upper wool layer can easily be removed and exchanged for new clean wool. The lower filter medium layers are rinsed out or replaced at greater intervals, depending upon the degree of dirt accumulation. Once the water has passed through the second chamber it arrives relatively clean into the third chamber. There the water is aerated and returned to the tank by means of an air lift system. Air lifts have great capacities for oxygenating water. The advantages of bio-chambers also include clear water, easy operation, and quick service access. Due to the large filter volume, the filter medium remains undisturbed for long periods of time, which facilitates maximum bacterial action to purify the water. The obvious disadvantage is that they are highly visible. Since this is not always desirable, it is recommended that the bio-chambers be appropriately camouflaged. Tanks with built-in

biological filter chambers are rarely commercially available, but they can be easily constructed before the tank is placed into service.

Very similar in function to biological filter chambers are trickle filters, which can also be used for supplementary filtration. They are, however, only marginally functional as the sole filtration for a cichlid tank; for instance for breeding. Basically, trickle filters function like biological filter chambers, but they are distinguished from the latter by having a larger filter volume and a slower flow rate. With few exceptions, trickle filters are rarely commercially available. Since trickle filters are not really suited for East African cichlids, I will omit a detailed description of them.

Suction filters can also be used as supplementary filters. They are rarely suitable as primary filters, since a sufficient water flow cannot be achieved. The intake tube of a suction filter should be affixed in such a manner that it does not pick up a lot of dirt particles. A suction filter should primarily be used to clean and recondition the water, since the bulk of the solid waste is already being removed by the

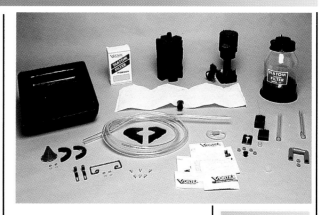

submersible filter pumps.

Irrespective of what type of filter system is being employed, it is important to select the filter with the proper capabilites. Or, more simply stated, the larger the filter the better the filtration effectiveness and the longer a filter

Vortex® makes filters with replaceable parts. This is a very important feature.

This filter pre-filters the water with a washable and replaceable filter sponge.

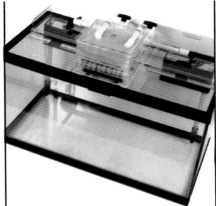

remains fully operational. You must not scrimp on the life-support system for your fishes. If in doubt it is better to select the next larger sized filter model. Two filtration systems operating independently from each other provide you and your fishes with a greater safety margin. It is well known that heaters always fail while you are on vacation. Filters also seem to always stop operating when you are on the first day of your vacation, and, in a similar vein, you always notice that your fishes are sick just after the corner aquarium shop has closed for the weekend. Properly functioning technical equipment is absolutely essential for the well-being of our charges and to your success as an aquarist.

BOTTOM SUBSTRATE

The decoration of an East African aquarium should be oriented toward elements that are similar to those of their natural habitat, which are, generally, sand, rocks, and very few underwater plants. First of all we need the substrate. You should select sand since this also occurs naturally in their native habitat. Moreover, cichlids like to "chew" (sift) sand with great perseverance. Some experts propose that this sand sifting is related to digestion, since sand is a common ingredient in the stomach contents of dissected East African Rift Valley cichlids. A definitive explanation of this behavioral phenomenon is yet to be given.

Some of the East African cichlids—predominantly

the bottom-dwelling forms—have proven to be highly sensitive to bacterial infections. Since a gravel substrate contains far more bacteria than ordinary sand bottoms, sand should be the preferred medium. Gravel, especially very coarse gravel, will quickly become unsightly with algae. You can, however, also use very fine gravel. But medium to coarse gravel must be regularly serviced, i.e., the debris must be siphoned off. Sand, on the other hand, will not retain any debris. Because of the smooth substrate surface provided by compacted sand, all debris, such as decaying plant parts, leftover food, fecal matter, and similar material will proceed unimpeded to the filter.

Another factor to be taken into consideration is the adaptability of cichlids. Depending upon the colors of background and

Lake Malawi habitat tank. Rock stuctures were assembled from limestone. Plants include *Anubias* species in the foreground, *Vallisneria* at the back.

substrate, the fishes will display lighter or darker color patterns, with weaker or more intense colors. The coloration of most cichlids becomes more effective with a lighter colored substrate, such as sand or light, fine gravel.

Sand and gravel are available in various grain sizes and different shades of color from aquarium shops and similar outlets. Sand can also be taken from some natural environment, provided this is permitted and the material is not contaminated with pollutants. It is important, however, that individual grains are smooth (i.e., without sharp edges). For instance, sand used in cement mixtures for construction purposes is unsuitable as an aquarium substrate. On the other hand, commercially available quartz sand can be used in an aquarium; but keep in mind that due to its near pulverized consistency it is easily sucked into the filter. Sand taken from a natural aquatic habitat has several advantages. It contains natural minerals and trace elements and it also has natural grains. Plants grow particularly well in such substrates.

Before the selected substrate material is placed into the tank it must be thoroughly rinsed. Some of the rinses should be made with hot water in order to kill undesirable bacteria. Gravel is most easily washed in a kitchen sieve or strainer, provided there

Fish nets are indispensable tools for the aquarist; those shown here are manufactured by Rolf C. Hagen Corp.

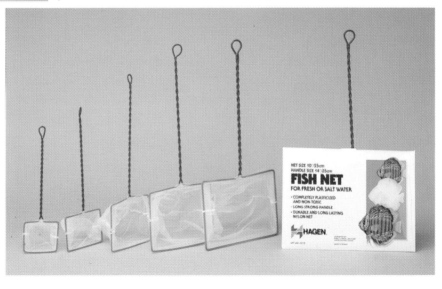

There is no question that living plants are best for cichlids and most other fishes. But cichlids can be tough on plants. They eat them and they tear them up. Plastic plants are the answer. Your local pet shop should have a wide selection. Use only plastic plants that have been designed for use in aquariums, such as the Second Nature Plantastics shown here. Plastic ornamental plants not designed for use in the aquariums may be toxic!

is one around that is not used for cooking purposes! A bucket is more suitable for washing out sand; it prevents excessive loss of sand with subsequent clogging of drains, etc. While adding water, the sand is vigorously stirred up in order to free-up dirt particles, which are then flushed out. Sand must be washed until the water remains virtually clear.

It is not advisable to place a layer of plant nutrients under the sand in a cichlid tank. Because of constant digging and sand sifting by the fishes, this

The filter is in place and so are the rocks. The branch coral has no place in a freshwater aquarium— and that's why the model is laughing!

would soon be uncovered and pollute the water. The relevant aquarium literature also often contains a suggestion that cichlid enthusiasts place a styrofoam layer between the aquarium bottom and the substrate above it. This is to prevent a rock from falling onto the glass bottom and breaking it. Usually, however, this styrofoam is uncovered by the fish and will then (much to the displeasure of the aquarist) float freely on the aquarium surface. Instead, it is better to plan for a thicker substrate layer, which then also buffers the fall of rocks. For safety reasons the substrate should be 5 to 7 cm thick.

All rocks available commercially or from natural sources can be used as tank decor except marble and metal-containing rocks (metallic inclusions). If you intend to keep epiphytic (aufwuchs) feeders you might want to give preference to smooth rocks, since these fish like to graze over the rocks in search of algae.

Calcareous rocks are common in nature. These are quite useful for aquarium purposes since they also "harden" aquarium water. Certain sedimentary rocks are particularly decorative, although environmentally conscious aquarists may wish to avoid using them.

This type of rock developed through the gradual erosion by seawater along the coast and it provides habitat space for many marine organisms. Since the large demand for this sort of rock has already caused substantial denuding along coastlines, further demands from aquarists should be curtailed.

Rocks must be arranged in such a way that their weight rests directly on the tank bottom. Carelessly assembled rock structures resting against the back wall of the tank can have

Place the full hood onto the aquarium.

dire consequences. The glass panel could shift from the weight, causing the tank to leak. In order not to affect overall tank stability, rocks should not be leaned against the back.

Rocks, by themselves, are already decorative elements, but keep in mind that cichlids like to dig. Therefore, all rocks should be resting firmly on the bottom. You can also build your own creative rock structures by gluing rocks together with silicone cement. However, the silicone cement must be permitted to harden properly and then be leached sufficiently. Quite often, a few larger rocks are more esthetically pleasing than a number of smaller ones. For larger cichlids it is advisable NOT to build any caves. Instead, arches

Make sure the aquarium's hood fits well.

There are many kinds and sizes of plastic plants. Use varied sizes so the taller ones can be used as backgrounds.

are more effective for these fishes. Small caves, on the other hand, can be constructed as hiding places for smaller fishes.

Caves are ideally suited for cave or cryptic substrate spawners. The size of the caves should correspond to that of the fishes that are to occupy them. But among mbuna cichlids, epiphytic feeders, and similar forms, the availability of caves leads to intensive territorial fighting. In this case, the dominant animals claim all caves for themselves and drive the other fishes away. Therefore, proceed on the basis that the fishes won't fight if you don't give them

anything to fight over. You then eliminate one element of aggression among your fishes.

Tree roots have no place in an East African cichlid tank. They are not normally found in the habitats of these fishes. On the other hand, far more suitable are some of the various ready-made caves that are available from aquarium shops. Using them in an aquarium is a matter of individual taste, since some of them tend to look highly artificial. Yet with some imagination and creativity these caves can be blended into the aquarium decor quite inconspicuously.

If you are going to use plastic rocks, get the kind that have caves, as the Malawi cichlids will appreciate a place to hide.

PLANTS AND PLANTING ARRANGEMENTS

Although there are few plants in the habitats of East African cichlids, you can certainly use some in your tank. Actual plant species from the respective habitats are rarely ever available. Yet some plants have proven to be quite suitable for these cichlid tanks. Particularly suitable are the *Anubias* species, as well as some *Cryptocoryne* species. These plants can take hard water as well as the abuse by cichlids.

All *Anubias* species can be used. They look particularly attractive on rocks, where they will eventually become established. But before the roots of *Anubias* are able to get a foothold you may have to secure the plants with a short piece of strong string or some lead weights that can be removed again later on.

Cryptocoryne usteriana and *C. balansae* are particularly recommendable from among the cryptocorynes. These species belong to the so-called "true" cryptocorynes, which means that they can be kept submerged throughout the year. In contrast, other species have to be kept temporarily at low water levels, otherwise *Cryptocoryne* fouling sets in.

C. usteriana and *C. balansae* are easy to keep. Moreover, the former grows

A specially planted patch of 'Lizard' *Cryptocoryne* is highly recommended. Juveniles like these Bulu Point *Tropheus moorii* will grow up together and form a harmonious group as adults.

Holger Windelov, the world's leading aquarium plant grower, developed this hybrid. He warns that cichlids might chew up any aquarium plant.

very rapidly and can produce individual leaves up to 80 cm long. This species is particularly well-suited for large aquaria. *C. balansae* is more delicate in appearance and grows somewhat more slowly. Both species of *Cryptocoryne* multiply through adventitious plants, that is, through runners.

Also suitable for an East African aquarium are species of *Vallisneria*, for simulating inshore reed banks. They, too, grow rapidly and are quite hardy. Anyone keeping predatory cichlids can also make use of less robust plant species, since these fishes do not feed on plants.

The above-mentioned plant species should be used predominantly for grazing cichlid species. Moreover, keep in mind that the natural habitat of these fishes contains only few plants, and there must be adequate open swimming space. A densely planted aquarium may be esthetically appealing, but it does not cater to the requirements of East African cichlids. Yet a few plants should be part of the decoration. Especially large plants contribute significantly to the removal of dissolved nitrogenous waste products from the tank water.

Plants with roots embedded in the substrate should at first be weighted down with rocks. This prevents them from becoming uncovered by the digging activities of the fishes before they have had an opportunity to become

A habitat photograph: Plant patches like this are rare in Lake Tanganyika.

firmly established in the substrate. Aquarium plants must be attended to on a regular basis. Dead and dying plant parts (stems, leaves, etc.) must be removed so that their eventual decay does not increase the bioload of the water. Moreover, they look unsightly.

AQUARIUM BACKDROPS

In addition to any interior aquarium decoration you should consider installing a complementary aquarium backdrop. These are available in a number of different motifs—including rock formations—from most aquarium shops. An aquarium backdrop rounds off the interior decoration of an aquarium. Living room wallpaper with obviously unnatural patterns are totally unsuited as aquarium backgrounds and they tend to distract the observer.

As an alternative to commercially available backdrops and patterns you can, of course, make your own. To do that you need thin styrofoam sheets of the same size as the back wall of the tank to be fitted, as well as a (children's) paint set. As a first step, the styrofoam sheet is cut to the exact size and the edges are cleaned off and smoothed. Then you start painting the styrofoam sheet with a mixture of predominantly white and some green and blue mixed in. That should give the styrofoam a uniform greenish-bluish color coating, which must be permitted to dry very thoroughly. Then you paint outlines of rocks and a sandy bottom onto the styrofoam very roughly. It is very important that precise contours and distinct color shades are avoided in order to achieve a natural effect. Once the paint has dried, the styrofoam sheet can be attached to the back of the aquarium.

This same process can also be done in water-insoluble oil paints. Such backdrops convey to the viewer a feeling of infinite depth in the aquarium. It is also ideal for photographic

Blue Ribbon Pet Products manufactures exciting aquarium backgrounds. One is shown here. Petshops have a wide variety of backgrounds.

purposes in order to achieve habitat-like photographs.

There is yet another way to use styrofoam as an aquarium backdrop. For that purpose we use sheets of at least 5 cm thickness or more. In order to achieve sufficient thickness you can also glue two or more sheets together, back-to-back, and then carve out the required features. This is best done with a hot soldering iron, to avoid crumbling of the styrofoam. You can carve out caves, depressions and crevasses, but the bottom layer of styrofoam must not be damaged. Having done all the required carving, the styrofoam sheets are then painted with water insoluble non-toxic paint to one's own imagination. The paint must be permitted to dry thoroughly. The completed backdrop is then attached to the inside of the tank with non-toxic aquarium silicone cement by applying the silicone over large patches on the back of the backdrop. Fitting and attaching the backdrop must be done in an empty tank.

Because of the inclusion of massive volumes of air in the styrofoam material, this sort of backdrop has considerable buoyancy. Therefore, it is important to

A closeup of a Blue Ribbon background.

apply sufficient silicone to achieve proper adhesion to the back of the tank. The silicone is then permitted to dry before water is added for a leaching period of several days. After that the tank is drained and it can then be decorated and set up as required. Although this method is very popular among aquarists, I personally cannot recommend it. After a while algae establish themselves on the backdrop and it becomes unsightly. Subsequent removal of the styrofoam backdrop is also difficult. In addition, there is the danger of flotation, because of the trapped air inside the styrofoam. A backdrop attached to the outside of the tank affords you the opportunity of removing and/or changing it at any time. Moreover, this arrangement makes it easier to service and clean the tank and to catch the fishes, if need be.

Before getting started you should figure out how much the entire set-up will cost and then get the largest aquarium setup possible within your budget. Many petshops have tank kits containing almost everything you will require to get started properly.

GETTING STARTED

Before we put any fishes into the aquarium it has to be thoroughly broken in. The tank is completely set up and all technical components are hooked up. Ideally, the tank should be kept fully operational and running (but without fishes) for 2 weeks. But impatient aquarists can expedite matters with the aid of bacterial cultures. These can be taken (as inoculating samples) from filter media that have been in use for some time in other tanks. If such material is not immediately available you may be able to purchase bacterial cultures from specialist shops or biological supply houses. Before the fishes are purchased all water quality parameters must be thoroughly checked: this includes the pH value, total hardness, and nitrite content. If everything is satisfactory the tank is then ready for the first cichlids.

The wattage of the heater you choose depends mostly on the size of your tank.

THE EAST AFRICAN RIFT LAKES

ORIGIN

As early as 1756 the German priest Theodor Lilienthal noticed that the opposing coastlines of various continents fitted into each other like parts of a giant puzzle, even though they were separated by oceans. Early this century (1912) meteorologist Alfred Wegener published his book *The Origin of Continents and Oceans* in which it was suggested for the first time that the ground under our feet is not stationary. Instead, it is in constant motion, so that the shape and form of the Earth is continuously changing. The technology of modern geology and paleontology enabled a comparison between the rock formations and provided convincing evidence of the former cohesion between Africa and South America. Although the cichlid species of these two continents are now separated by the Atlantic Ocean, they are still of the same origin.

Forces within the Earth have separated the globe into so-called continental plates. The African continent is situated mostly on the African continental plate, which extends into the Atlantic up to the Mid-Atlantic Ridge. But a small part of Africa is located on the so-called Somalian continental plate. Because of the drifting apart of these two continental plates, the East African Rift Valley and, eventually, the lakes, developed. Earthquakes along the rift valley fault line as well as volcanic activities are evidence of the continuing movements within the Earth. The fusion line of the continental plates displacement extends into the Israeli-Palestine Jordan Valley via the Red Sea, and continues via Ethiopia to the Indian Ocean (Straits of Mozambique). All lakes along this fusion line are referred to as Rift Valley lakes. They start with Lake Rudolf south of the Abyssinian Highlands, and continue southwestward

with Lake Albert, Lake Edward, Lake Kivu, Lake Tanganyika, and, finally, at the southern end, Lake Malawi. In its full length the Rift Valley is more than 6,400 km long and its origin dates back about 17 to 18 million years.

Lake Victoria, located somewhat off to the side, is not considered to be part of the Rift Valley lakes. As already indicated by its shallow water depth, its origin is due to an elevation of the surrounding areas forming a shallow depression that subsequently became filled with water. This elevation was due to continental drift, whereby the area immediately adjacent to the fissure line was pushed away and so became elevated.

The very deep Rift Valley lakes developed in the fissure formed as the continental plates drifted apart. The surface of Lake Tanganyika is 773 m above mean sea level, but because of its depth of 1,423 m another 650 m are below sea level. This is also the lowest point on the African continent. Lake Tanganyika is the second deepest freshwater lake on Earth, after Lake Baikal, which also originated from a rift valley formation. Lake Tanganyika is estimated to be about 10 to 15 million years old and Lake Malawi about 2 million years old. A further drifting apart of the continental plates in Africa could one day possibly lead to penetration by salt water. This could conceivably occur from the Red Sea, but also possibly from the Indian Ocean.

DISCOVERY

European interests in Africa were awakened by the expeditions of the Portuguese. In the foreground was the perpetual question as to the source of the Nile River. The desire to solve this mystery was the prime motivation for all journeys of discovery into Africa up until the end of the 19th Century. Research into Africa got its most significant push in 1788 with the establishment of the "African Association," the precursor to the "Royal Geographic Society," which was founded in 1830. Without this, the exploration of Africa had to be viewed as being less for scientific reasons and more to satisfy political and economic interests. Following the exploration of the Niger River and that of the Sahara Desert, East Africa became the focal point of interest from about

Facing page: Pseudo-tropheus zebra."

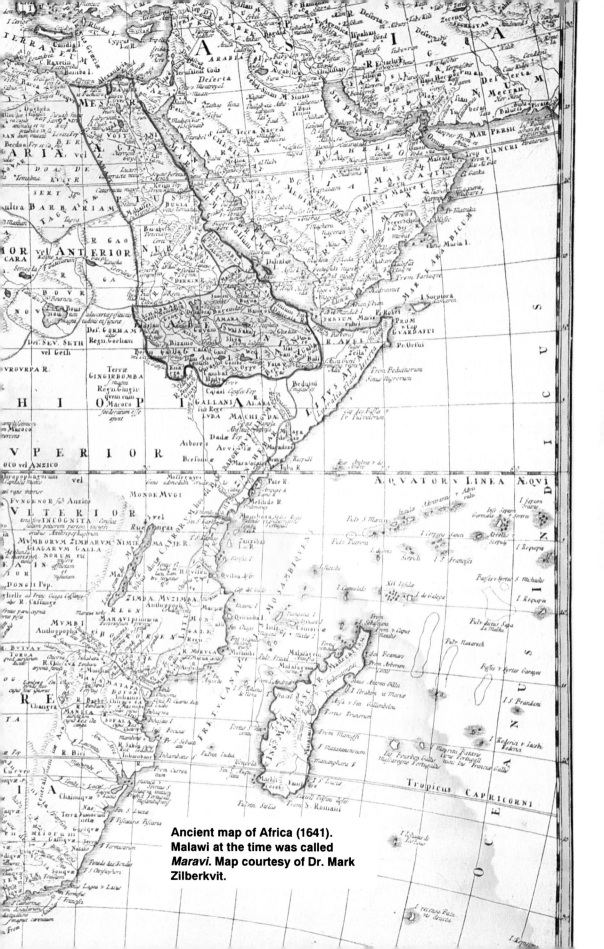

Ancient map of Africa (1641).
Malawi at the time was called
Maravi. Map courtesy of Dr. Mark
Zilberkvit.

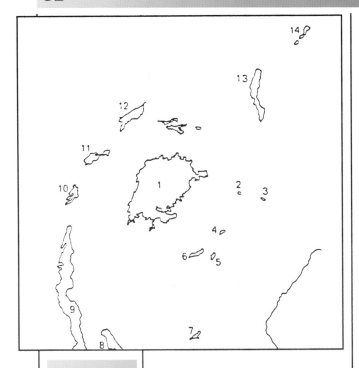

Satellite view of the Rift Lakes of Africa. Photo courtesy of the National Aeronautical and Space Agency (NASA).

(1) Lake Victoria

(2) Nakura

(3) Elmenteita

(4) Natron

(5) Manyara

(6) Eyasi

(7) tip of Lake Malawi

(8) tip of Rukwa

(9) tip of Lake Tanganyika

(10) Kivu

(11) Edward

(12) Albert

(13) Turkana

(14) Abaya

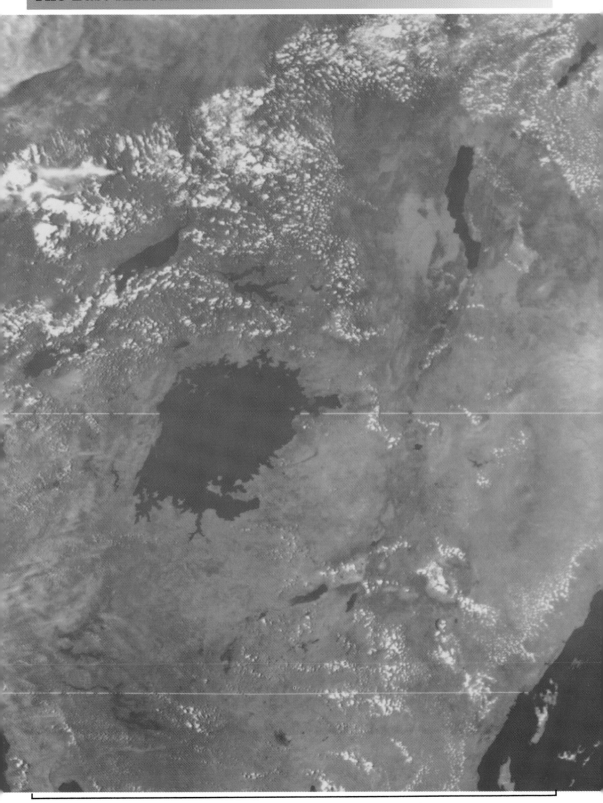

1850 onward. From Mombasa, in 1848, two German missionaries, the Brothers Redman, ventured into the interior. However, their activities created uneasiness for the British Crown, since both missionaries openly advocated "guardianship by Christian nations." In conjunction with this there were demands for a colonization of this area. These then were the early delineations for the later subdivision of Africa among the European powers. The expeditions that followed have to be judged against these political backgrounds.

The most exemplary efforts in exploring East Africa have to be contributed to the Scottish missionary David Livingstone. He travelled to East Africa at the age of 27 and for many years journeyed alone through the region of the East African lakes until his death in 1873. He died at the age of 60 from the effects of a disease.

After the discovery of the East African lakes, zoologists and botanists followed to explore the lakes in detail. Ichthyological research by numerous scientists followed.

LAKE VICTORIA

From among the many East African lakes only cichlids from the Great Rift Valley lakes have generally found their way into the tropical fish trade. From among these, species from Lake Malawi and Lake Tanganyika have become particularly popular. Lake Victoria is already some way behind. Covering an area of 68,000 km², Lake Victoria is the largest of the East African lakes. In size it corresponds roughly to the area of Bavaria and is the second largest freshwater lake on earth. Among the three large East African lakes it is geologically the youngest, with an estimated age of about 1 million years. From among the many cichlids occurring in Lake Victoria, species of the genus *Haplochromis* form the largest component. Unfortunately, there are very few regular tropical fish imports from Lake Victoria due to the absence of local fish exporters. Although the lake's fish fauna is still largely unexplored, the species of *Haplochromis* in particular—about 300 endemic species—are already threatened with extinction. The Nile Perch, *Lates niloticus*, has invaded Lake Victoria from hatchery

Facing page: An Aerial view of Domwe Island, Lake Malawi.

The smallest of the slender cichlids, *Julidochromis transcriptus*. It prefers caves and crevices.

ponds located at the northern shoreline. This event preempted an official release of Nile Perch into the lake. The predatory Nile Perch found an excellent food source in the endemic fish fauna, which consists predominantly of haplochromine cichlids. This led to an alarming increase in Nile Perch, and at the same time a corresponding severe decline in the cichlids. This now precludes a complete ichthyological investigation into the original fish fauna of Lake Victoria, since some species will have already become extinct.

The Nile Perch was originally supposed to be deliberately released into Lake Victoria in order to become a fat and protein source for the people of the area around the lake. But the rather oily Nile Perch cannot be sun dried (like the haplochromine species). Instead, in order to cure and preserve the meat Nile Perch must be smoked. This in turn has led to an increased demand for wood, and so there has been progressively advancing denuding of the forests around the lake. In addition, catching Nile Perch requires special nets, which the majority of the people cannot afford. Consequently, catching Nile Perch has become restricted to those who are financially already better off. The previously fished for *Haplochromis* species, formerly occurring in huge numbers, could be caught in simple nets.

The relatively young age of Lake Victoria leads one to the conclusion that the specialization of cichlid species has not yet advanced as far as in the other two large lakes. Therefore, the predominantly occurring *Haplochromis*-like cichlids often exhibit only minor deviations in their external appearance. Consequently, species identification is often very difficult. Since Lake Victoria has few rocky areas, most endemic cichlid species have adapted to a life in the open water above a sandy bottom. The largest number of species are opportunistic feeders, as a consequence of the available habitat. Their diet includes eggs, embryos, larvae, and juvenile fishes. A small number feed on crustaceans, snails, plankton, algae, and higher plants. Unfortunately, live fish imports from Lake Victoria have been rather sparse, for the reasons outlined above.

LAKE MALAWI

The age of Lake Malawi is estimated to be several million years. This makes it older than Lake Victoria, but younger than Lake Tanganyika, with an estimated age of 10 to 15 million years. Lake Malawi (formerly Lake Nyasa) has been isolated for about 2 million years. It is 600 km long and in some parts up to 80 km wide. The surface of Lake Malawi is 472 m above mean sea level. Its maximum depth is 700 m. The only runoff from Lake Malawi is via the Shire River, which flows first into

The rocks on the shoreline of Lake Malawi are identical to the rocks under water. The fishes hide in the crevices and interstices among the rocks.

Lake Malombe and then continues on into the Zambezi River. The latest research indicates that 25,000 years ago the water level of Lake Malawi was about 400 m below its current level. In contrast to Lake Tanganyika, however, Lake Malawi remained undivided during its low water level. Increasing adaptation to the environment occurred particularly through specific changes in the mouth and dentition. This facilitated variable food intake and feeding specializations. The current inventory of cichlids in Lake Malawi is estimated to be in excess of 600 species.

In nearly all locations, the water in Lake Malawi is crystal clear, and underwater visibility can be up to 20 m. Individual shoreline segments of the lake consist of sand banks and rocky areas. Most cichlid species of interest to aquarists inhabit the rocky littoral zone. Cichlids occur principally in the shallow water zones; their numbers decline slightly below a depth of 10 meters and fewer species are found below 20 meters. The densely inhabited rocky littoral zone has, on

A view from the garden of the Peter Davies homestead in Cape Maclear, Lake Malawi. The name of his boat is *Mbuna*, as you can see.

average, 6 to 7 fish per square meter, sometimes even up to 12 fish. Such dense occupation is facilitated by the great availability of suitable food. The heavily algae-covered rocks accommodate multitudes of small organisms. A sample count of one square meter of epiphytic growth revealed that up to 30,000 organisms can live in such a small area. These include mainly ostracods, chironomids, and copepods.

Cichlids living in these rocky areas are characterized by specific requirements and certain behavioral peculiarities. They are called "mbuna" by local fishermen, a term since adopted by aquarists. The term mbuna refers predominantly to all *Pseudotropheus, Labeotropheus, Melanochromis, Labidochromis, Iodotropheus, Gephyrochromis, Genyochromis, Cynotilapia* and *Petrotilapia* species. Most mbuna attain a maximum length of about 10 cm, with only a few of them getting to be 15 to 20 cm long. During the breeding season, which is not tied to a particular time of the year, the males defend small territories. Mbunas have a strong (permanent) territorial instinct wherein the territories are rarely ever left even in the pursuit of a rival. When endangered, mbuna will quickly flee into the nearest crevice, but even then the territory is not abandoned. And so even clearly delineated sandy areas will, in this manner, form an insurmountable distance for mbunas.

Following the rocky littoral zone there is a transition zone of sandy and rocky habitats. In large intervals the sandy bottom is occasionally covered with stands of *Vallisneria* and sometimes also hornwort (*Ceratophyllum*). Representatives of this transition zone include *Aulonocara, Hemitilapia, Oreochromis, Nimbochromis*, etc., as well as juveniles of other cichlids. *Aulonocara* is divided into two species groups. Most commercially available species belong to the group of smaller *Aulonocara*, which also includes the more colorful species. The *Aulonocara* of the transition zone move into "sleeping" caves at night. As in many mbunas, mating also takes place inside a cave. During the day edible material is "chewed" out of the sandy substrate. Even fine sand contains large numbers of tiny crustaceans, mosquito larvae, copepods, and especially ostracods. In addition, *Vallisneria* stands

An underwater scene in Lake Malawi showing the cichlids feeding on the algae and the invertebrates in the algae.

Above: A view of Monkey Bay, Lake Malawi, from David Eccles's photo. **Below:** The fishes are fairly well camouflaged in the tall grasses in Monkey Bay along the shore of Lake Malawi.

The shallow underwater shelf along the shore of the non-rocky coast of Lake Malwi. Most of the coast is rocky.

contain snails and mussels, and a shrimp species, as well as various insect larvae. The transition zone and the sandy littoral areas close to shore also accommodate species of the genera "*Haplochromis*" and *Lethrinops*. But these are also commonly found in open water. With approaching danger they attempt to escape in wild flight. "*Haplochromis*" *rostratus* can also bury itself in the sand when endangered.

Another habitat is characterized by reed stands, just as in the sandy littoral zones and in the transition area to the rocky littoral zone. These reed patches serve as breeding grounds for birds, but at the same time they also provide hiding places for cichlids against predation by birds. None of the mbuna species were found to be endemic to these reed areas. "*Haplochromis*" *compressiceps*, on the other hand, appears to have become most closely adapted to the reed zone. This species, which grows to 25 cm in length, lies in wait in these reed patches for unsuspecting prey. Juveniles of various *Oreochromis* species can also be found here,

sometimes forming large schools.

Deeper water accommodates the "utaka" cichlids, a term also adopted from local fishermen. 'Utaka' refers to a group of about 20 "*Haplochromis*" species (for example *H. mloto* and *H. flavimanus*), that feed by filter feeding on zooplankton. These fishes also enter inshore habitats in search of food. Other typical representatives of the deeper open water zone are *Rhamphochromis*, with about 8 species. These cichlids, which can grow up to a size of about 60 cm, are exclusively predatory.

Whereas nearly all the known cichlid species of Lake Malawi come from this lake's southern and western coastline, the east coast has so far remained largely unexplored until

recently. There are regular imports of Lake Malawi cichlids by local aquarium fish exporters. But the fish exported are only a small fraction of the total fish catch of Lake Malawi. Most fishes caught are utilized as food for the local population.

Tropheus brichardi from Cape Katende, Zaire.

The photo above shows a Lake Tanganyika habitat. Compare it to the photo here taken in Lake Malawi.

Lake Malawi. When the lake is churned up it looks more like ocean surf.

LAKE TANGANYIKA

Because of its size and depth, Lake Tanganyika is the most voluminous of all the East African lakes. Its length is 670 km and its width is up to 80 km. As has already been observed in Lake Malawi, the water level in Lake Tanganyika has repeatedly changed. At its lowest level the entire lake was divided into two larger basins and one smaller one, which were separated from each other by a subterranean mountain range. To this day there are southern and northern races and forms among some of the cichlid species.

Specialization among cichlids in Lake Tanganyika has become particularly far advanced. It is hypothesized that the reasons for this phenomenon are, on the one hand, the greater age (in contrast to the other two large lakes) with a longer period of isolation, and, on the other hand, the periodically lowered water level with a concurrent separation among individual species. This has contributed to the development of new species and subspecies.

Lake Tanganyika has a tributary in the north, the Ruzizi River, which drains Lake Kivu. The largest tributary in the east is the Malagarasi River, which discharged into the Congo Basin prior to the

supposed to have entered Lake Tanganyika via the Lukuga Trench, but their distribution is restricted to the immediate proximity of the Trench, near Kalemie.

In Lake Tanganyika, just as in Lake Malawi, the various coastal sections tend to alternate. They are characterized by rocky shorelines and sand banks. Those cichlid species of

The shore of Lake Tanganyika.

development of Lake Tanganyika. The split-off section of the Malagarasi River in the west of the Lake is now referred to as the Lukuga Trench. It is not a tributary, though it does form a connection between the Congo Basin and Lake Tanganyika. Various fish species are

interest to aquarists occur predominantly in the littoral region close to shore. Here too, the rubble zone is the habitat with the densest fish fauna. This zone consists of rocks and boulders of different sizes. Most of the cichlids occurring in this region are mouthbrooding grazers.

Specialized grazers (= epiphyte or aufwuchs feeders) can readily be recognized externally by their subterminal mouths, special dentition, or slightly open lips. This is the habitat of the cichlid genera *Eretmodus, Tanganicodus, Spathodus,*

Neolamprologus tretocephalus and *N. sexfasciatus,* are specialized feeders on snails, mussels, and small crustaceans.

The transitional zone between the rocky and sandy littoral zones accommodates

Underwater view of Lake Tanganyika.

Tropheus, Petrochromis, and *Simochromis.* In addition, there are cave spawners, such as *Neolamprologus, Julidochromis, Telmatochromis,* and *Chalinochromis.* These are small to medium-size predators that frequent crevices and also deposit their eggs there. With their small, sharp teeth they grasp eggs, juvenile fishes, and shrimp as well as various insect larvae. Some cichlids, for example

Ophthalmotilapia, Paracyprichromis, and *Cyprichromis* species. Larger grazers, such as *Tropheus, Petrochromis,* and *Simochromis* also occur here. In the sandy littoral zone we find the so-called "sand fishes." These include *Callochromis, Enantiopus, Grammatatria, Aulonocranus,* and some *Xenotilapia.* The latter, however, also occur in the sandy littoral zone close to shore, since they are largely migratory. Other

Above: A view from a hill looking down onto Lake Tanganyika.
Below: An underwater view of Lake Tanganyika.

Such rocky shores continue under water and provide the habitat for many cichlids in Lake Tanganyika.

A view along the coast of Lake Tanganyika.

cichlids of the deeper sandy littoral zone include the many species of snail cichlids. They belong to the genus *Neolamprologus* and occupy empty shells of the aquatic snail *Neothauma*. The sandy littoral zone also has sizable stands of higher plants such as *Vallisneria* and *Myriophyllum*. These are nearly totally absent from the rocky littoral zone.

In deeper water (from 10 to 20 m depth) we find *Cyphotilapia frontosa* and other predatory species. They occur, however, always close to small rocky reefs in the sandy littoral zone. The open water is inhabited by other, mainly planktonic feeding cichlids of the genera *Benthochromis*, *Trematochromis*, *Paracyprichromis*, and *Trematocara*. Since the drifting food for these fishes tends to sink to greater depths during the day, the fish will follow their food downward. The deeper water also accommodates the most aggressive predators among the cichlids, species of *Bathybates* and *Hemibates*. Because of their large size they may not be suited for home aquaria. The west coast of Lake Tanganyika, which is located in Zaire, remains largely unexplored. Therefore, one would expect additional new species from that region of the lake.

In back of Cape Maclear is this rocky terrain. Compare it to the Tanganyika terrain on the facing page.

In Lake Tanganyika sandy patches alternate with the rocky shore. A fishing village along the shore.

THE CICHLIDS

AGGRESSION

In order to be able to keep East African cichlids successfully it is necessary to take a look at their natural behavior. This behavior is determined by food requirements, reproduction, and species-specific behavior. In the first instance, any natural aggression is directed toward feeding competitors. This includes sibling fish as well as other similar-looking cichlids. The outward similarity of various cichlids is nearly always based on identical food requirements. Competition for food is expressed by stronger fishes driving off the weaker ones. Depending upon specific food requirements, a more or less sizable territory is claimed within a particular habitat. The establishment of territories not only serves to secure an adequate food supply but helps assure reproduction as well. It is not uncommon to find a fish density of 8, 10, yes, even 12 individuals per square meter in some parts of the the East African lakes. This is facilitated by variable food intake, reproductive behavior, and external appearance. On the other hand, food competitors will also defend the territory required for their own survival against each other as well. Personal observations and experiments have shown that aggression within the aquarium is also influenced by food availability. While aggression increased with an insufficient food supply, a group of cichlids behaved distinctly less aggressive when food was abundant. I shall refer to this point again in the chapter on feeding.

Another type of aggression develops toward siblings of the same sex. Every fish attempts to reproduce; nature provides the selective processes. Dominant animals are better adapted for survival than weaker ones, therefore they can breed more often. Consequently, a sibling of the same sex is not only a food competitor but also competes for reproductive opportunities. For the latter there is a severely aggressive response in cichlids. The competition for such opportunities leads to severe fighting not

Fish collecting and holding station in Zambia. Cichlids are sent all over the world from there.

only among males (as is often assumed), but this also occurs among females. If there are two females ready to spawn in the proximity of a courting male, each female will attempt to drive off the other. On the other hand, if a female ready to spawn swims into the territory of a single male, the male will initiate courtship behavior. If, in turn, the female signals—through shaking or trembling of the body—that she is willing to spawn, she is allowed to remain in the territory where the courtship continues. It does not matter if the female initially appears somewhat "reluctant." While the male continues courting the female, which in turn continues to demonstrate her willingness to spawn, she may also start looking for food in the male's territory. In this case food-dependent aggression is being suppressed by the reproductive instinct. A totally different sequence of events takes place if the female proves to be unwilling to spawn. In such cases the male sees only a food competitor, which is then relentlessly driven off. The same behavior can also be seen in a breeding pair as soon as siblings enter the breeding territory. These events can be observed commonly in an aquarium.

The reactions among tank inhabitants toward each other are far more intensive under captive conditions. Therefore, it is important to select species carefully so that a harmonious cichlid population becomes established in a home aquarium. The criteria to achieve this are essentially proper species selection, specimen numbers, sex ratios, and the size of the tank. In terms of space availablity we cannot

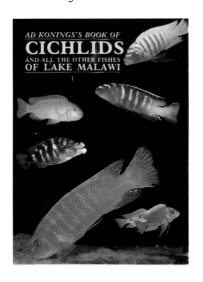

duplicate natural habitat conditions in an aquarium, so, suppressed males, rivals, and females unwilling to spawn cannot leave the territory of the dominant male as easily. If you attempt to do this anyway, some fish may be forced to jump out of the tank and others will be killed. Careful selection and stocking of the aquarium will make it possible to minimize aggression.

When buying your fishes you must also keep in mind that cichlids will establish a hierarchy. In East African cichlids we have highly developed, intelligent fishes. Social structures develop that are defined not only by the above-mentioned factors but also

Your local petshop has large books with thousands of photos in each one to help you identify the cichlids from the Rift Lakes of Africa including Lakes Malawi and Tanganyika.

by variable characteristics from within the particular cichlid species. These characteristics parallel human traits. Among cichlids there are those that "show off" and those that are timid, peaceful ones and aggressive ones, those that are deceitful as well as loners, to mention only a few of the characteristics that can be seen. Cichlids serve as study objects for ethologists to be used as inferences for human behavior. For aquarists it is often helpful to turn this procedure around. Here is an example: Just imagine one man and three women have become stranded on an isolated island. It does not take much imagination to picture the inter-human relationships that would develop under these circumstances. The male will quickly feel desired and being cared for; after all, he is being spoiled by his three women. On the other hand, the relationships between the three women will proceed along totally different avenues. Each will endeavor to become

Rubble zone in Lake Malawi.

elevated in the esteem of the man. This is the start of jealousies and fighting among the women. No doubt you will now say that fishes are not humans.

You are quite right, there is a big difference. Humans have learned (with some exceptions) through the use of appropriate, tactful words to more or less get along with each other. Fish do not have this capability. They attempt to express themselves through the language of signs and gestures. If this is unsuccessful, there is only physical escape, if possible. If you now transpose the above example to aquarium conditions, you will notice that the situation in the aquarium is the same as that on the island.

Having reached this point now, you may well wonder whether you still want to keep cichlids or, rather, some other fishes. Certainly, cichlids are not exactly uncomplicated pets. But take it from me, that is exactly the fascination of cichlid care!

SPECIES SELECTION CRITERIA

In this section I would like to provide some specific guidelines for selecting appropriate species combinations for

stocking an East African cichlid tank. The best method in order to establish a functioning cichlid community is to select fundamentally different species. The more the individual species differ in external appearance and requirements from each other the better it is. Examples of this are listed in the following chapter. I particularly recommend starting out with juvenile fishes. If you are starting up a tank completely new and all fishes are juveniles, they have more of an opportunity to adjust to each other. The inevitable hierarchical fighting among juveniles usually remains more or less subdued and certainly never ends with fatalities. This, then, eventually leads to a well

Typical representatives of the slender cichlids: *Julidochromis regani* is in the background while in the foreground are juveniles of *J. regani* and *J. transcriptus.*

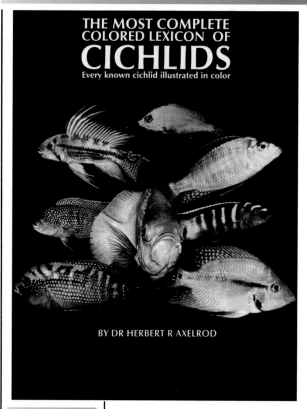

THE MOST COMPLETE
COLORED LEXICON OF
CICHLIDS
Every known cichlid illustrated in color

BY DR HERBERT R AXELROD

The most complete book on cichlids ever written is Dr. Axelrod's *Cichlid Lexicon*.

adjusted and harmonious group of cichlids, and losses due to severe hierarchical and territorial fights as adults are significantly reduced. This is particularly important if there is to be intraspecific harmony later on.

The number of specimens per species is dependent upon the size of the tank and the type of reproduction involved. For instance, mouthbrooders do not usually require fixed territories in the tank. Since the females pick up the eggs with their mouth during mating and keep them there until well after they have hatched, the female is not tied to a territory.

Things are somewhat different with substrate (cave) spawners. Since the eggs and later on the embryos are hidden, an active territorial defense by the broodtending pair is unavoidable. The demand for space is therefore greater.

For monogamous (permanently pair-bonded) cichlids it is advisable to purchase at least 4 to 6 juveniles or a firmly bonded adult pair. A mated pair will eventually develop from within the group of juveniles. The remaining siblings will be driven off as soon as courtship and mating begins and the need for establishing a territory arises. If the tank is large enough the remaining fish can withdraw into other territories. If the tank is too small for that the unmated fish should be removed in order to protect them from serious injuries. Once a firm pair has crystallized out of the group of former juveniles, such a pair is often permanently bonded. Many monogamous cichlids are rather selective when it comes to picking a partner. Often it is not enough to simply put a male and a female together in order to get a breeding pair.

Perhaps even among cichlids there is such a thing as "...love at first sight...!" Pairs that have become bonded in this manner are far more stable than those where the sexes have been placed together at random.

Pair bonding can be intensified through the presence of "enemies." In a community tank a cichlid pair will jointly defend its breeding territory against suspected enemies, e.g., against other cichlids. This joint defense effort tends to intensify the bond between the partners. When the "enemy factor" is absent, as is, for instance, the case in a breeding tank, aggression may be directed against the partner. Since the aggression cannot be discharged against an enemy, it is often the weaker female that then provides the required aggressive relief by being attacked by the male. Consequently, the female is driven off as soon as she indicates she is not ready to spawn. Normally, the

Center of photograph: A threat display by *Tropheus moorii*. All fins are spread to impress a female or to intimidate a rival.

Petrotilapia tridentiger, one of the mbuna, is a highly specialized epiphyte feeder (grazer).

female would then leave the territory of the male. This, however, is not always possible due to a lack of space in the tank. So the female remains in the tank and becomes a constant provocation for the male. Once this sort of situation has arisen, it is time to remove the male. It would be wrong to take the female, the weaker animal, out. Should the female have to be returned later on, the already stronger male would have the added "home advantage"; after all, it is his traditional territory.

This method works better the other way around. Once the male is gone, the female can recover on a good diet and start to develop eggs. When the male is returned to this tank it then becomes the intruder, but because of its physical superiority it will invariably establish itself again as the territory holder. By then the female will be ready to spawn and so the relationship stabilizes itself again. If, on the other hand, things do not work out as expected, it is advisable to transfer the pair to a larger tank with enemy factors. For this purpose the female is placed into the new tank a few days before the male in order to strengthen her position. It is also advisable to wait until the female is actually ready to spawn again.

Separating the pair must also be considered when the female has to remain in hiding and so cannot get any food. She quickly loses condition and is then in no position to start developing eggs again. When she tries to leave her hiding place to look for food, she is driven off again by the male, and so a vicious cycle develops. The female is not tolerated in the male's territory because she is not capable of spawning. On the other hand, she cannot develop eggs because she cannot get any food. When this happens, the time has come to help the female by removing the male. If it is impossible to reestablish this relationship, one of the partners should be exchanged permanently.

Matters are somewhat different when keeping polygamous cichlids. They are best kept in a larger group, although experienced aquarists can also keep them in pairs, but it does require a certain expertise. Keeping these cichlids in pairs is best done in a community tank where they are kept together with other cichlid genera. For instance, a tank full of *Neolamprologus tretocephalus*, *Julidochromis marlieri* and *Ophthalmotilapia ventralis* can also quite easily accommodate a pair of *Tropheus duboisi* or *Tropheus moorii*. Here, too, aggression will be minimized by the presence of the other aquarium inhabitants. However, there is no assurance that this will always work. Keeping monogamous cichlids in pairs should only be done according to the guidelines provided above.

For the permanent care of polygamous cichlids it is much simpler to acquire a sufficiently large group of animals, with preference to be given to juveniles. Juveniles are less expensive so you can purchase more individuals of one species and you will get even more pleasure out of watching them grow up. This does, however, require a bit of patience waiting until they have matured and begin to show their full adult coloration. The group should consist of at least 10 specimens with the ideal group consisting of about 15 fish. The larger the group, the smaller the intraspecific aggression, which, after all, is particularly strongly developed. Since there are numerous color variations

Threat display by an adult *Neolamprologus ocellatus* male. Males are easily recognized by their orange colored dorsal fin border.

among some of the polygamous cichlids, such as *Aulonocara, Tropheus*, and others, this decision has to be made with careful deliberation. A collection of the various color varieties should be avoided. Apart from the problem of possible crossbreeding, the number of individuals per group also becomes too small. Restrict yourself to a few color varieties of which you are particularly fond and where interbreeding is not possible.

As already mentioned, intraspecific aggression is greatly reduced in larger groups. This is due to the fact that in a large group individual animals can no longer pursue other siblings as undistracted as when they were in a smaller group. The hunted are easily lost and out of sight of the hunter, since too many other fish cross its path. The pursuit is then quickly terminated. Such a group of cichlids establishes a hierarchy, subject to certain criteria. This hierarchy, for instance, does not apply identically to both sexes with females as well as males having their own hierarchy, and the entire group forming a third hierarchy. Position within the hierarchy is determined by physical strength, perseverance, assertiveness, and sex.

The hierarchy within either sex is principally based on size. It is rare to find smaller, weaker fish of higher status than larger fish due to more aggressive instincts of the former, but this is common in the overall (group) hierarchy. Although here, too, the order is determined largely by size and strength, the males are clearly disproportionately better represented in the upper end of the order. It seems that males are equipped with a distinctly more intensive aggressive instinct than females. Consequently, even subadult males can win out over clearly larger females. When the fish are kept only in pairs, you can take advantage of these characteristics. Ideally you will want to pair off an at least identically-sized female with a male. I was particularly successful with this method. Many good cichlid "marriages" have started that way! It is important to make sure that the male is not clearly younger than the female nor should he be too small. Otherwise, it is quite possible that the male will not be accepted by the female.

But now back to keeping

cichlids in groups. Anyone who has ever observed young cichlids during the first few days of their (free-swimming) life will have noticed that even then the first hierarchical positions are already being established. Barely released from the mouth of the female, the tiny fish start to fight for dominance. It is quite funny to watch them as they pull out all stops from the entire range of behavior, in just the same way they will do it later on as adults. Rivals try to impress each other with various threat displays and then chase each other around in the tank. Yet, fatalities do not normally occur until the fish have reached sub-adulthood. The chances for a harmonious group of cichlids are significantly increased when starting out with juveniles.

It is, of course, important here that the group has grown up together. It is not advisable to put together a group of sexually mature fish originating from

Aulonocara steveni, one of the smaller and more attractive representatives of the genus.

Neolampro-logus ocellatus mating. The female deposits her eggs inside an empty snail shell, the male subsequently fertilizing them. The female inside the shell shows a distinct lateral band. This is characteristic for ocellatus *females which are spawning or involved in brood care.*

different tanks. In such cases, problems are often encountered. New hierarchical positions will have to be established, often followed by severe fighting, which may even lead to fatalities.

It is often erroneously assumed that there can be only a single male within a group of polygamous cichlids. This is not true! For instance, in the natural habitat *Tropheus moorii* can be seen in groups of about 20 individuals of which half are males. This sort of condition can also be maintained in an aquarium.

Basically, problems arise with groups where there are only a few males. Let us take a look at the various scenarios that can occur. If a male is kept together with a small or larger group of females, the male will be accepted by all females. Such males have no competition. Since a

second male, which could occupy the second highest position in the hierarchy, is absent the strongest female is elevated to that position. Females will compete for that position. This female will then assume, sometimes in conjunction with other females, a male-like behavior. Frequently such females will attempt to court other sibling females. This leads to instability within the group, which is accompanied by constant fighting and frequent chases through the aquarium by females. If the group of females is small (consisting only of a few individuals), usually none of the females will mouthbrood. As soon as a female carries eggs in her mouth, she is attacked by one of the other females. The female then either swallows the eggs or spits them out; in both of these cases they are effectively lost.

Within larger groups such aggression among females is less intense, and there the strongest females usually will carry their eggs to full term. The dominant female will then leave the rest of the females alone.

In smaller groups the male is usually very aggressive toward those few females present. Due to the small number of

individuals the male's aggression is spread over fewer animals. Too much stress for the females also means that the animals are not feeling well and there is also lack of brood care. Simultaneously, all cichlids in the small groups, except for the dominant animal, display little or no coloration at all.

If we put two males into a tank that is 1.5 m long, then there is sufficient space for two territories, provided, of course, both males are about equally strong. In this case, females do not have a high ranking in the hierarchy; the male-like behavior does not occur. The aggression of one male is directed primarily toward the rival in the other half of the tank. Generally, however, both males respect each other's territory. Occasional threat displays in the territorial border region make the point: "Here I am, this is my territory." If the tank is too small, keeping two males together becomes problematic.

If we keep several males within a group of cichlids, the size of the group plays an important part. For instance, with a ratio of 3 : 7 the group is too small for all males (unless you have a tank of more than 600 liters volume and there are

also other cichlids in the same tank). For each male one has to calculate a territory of about 50 to 70 cm (continuous) tank length. This can be reduced if the group consists of 15 animals or more. In this case excess males without territories have an opportunity to swim within the female group. Because of the size of such groups it is difficult for the dominant male to pursue any of the other males. Usually any other male entering the territory is simply driven off back into the female group.

The advantage of keeping two or more males lies in the fact that aggression is primarily directed toward rivals. At the same time, females have a natural

This pair of *Altolampro- logus compressi- ceps* has become bonded within a group of siblings. The female takes over brood care while the male defends the territory. When servicing the inside of the tank the aquarist's hand is often attacked by the male.

Neolampro-logus ocellatus female with young. The female has already developed eggs again. In contrast to the male, the dorsal of the female has a white margin.

opportunity for mating. In nature, females of polygamous cichlid species ready to spawn enter the territories of several males for breeding purposes. This facilitates the passing on of diverse genetic material. In an aquarium situation we can also achieve this situation with two or more males.

There is another important factor favoring the maintenance of larger groups. Since aggression between individuals in a larger group is substantially reduced, none of the animals feels particularly threatened or pursued. The general well-being of the group thus assured, the animals will respond by displaying full coloration. Harassed animals will not display their natural coloration.

The above-mentioned behavior patterns apply especially to very aggressive cichlid genera. These include to a large extent specialized grazers. *Aulonocara*, among the polygamous cichlids, can be considered as a relatively peaceful genus. Keeping these fish in pairs and together with other East African cichlids does not present a problem. The number of *"Haplochromis"* species is quite large, and the habitat and food requirements among these cichlids are highly variable. Consequently, a definitive statement about the aggressive instincts in these fishes cannot be made.

A number of cichlids are inclined to display their full colors only intermittently. This applies, for instance, to species of *Aulonocara, "Haplochromis"* and *Ophthalmotilapia*. If some of the species within these genera are being kept in pairs, the males will rarely ever display their full, magnificent coloration. It is only shown if there is a rival or a female ready to spawn in the same tank. For these reasons, it is advisable that they be kept in groups only.

The East African cichlid tank should not become a random collection of fishes. Confine yourself to a few cichlid species that have different requirements. This will assure not only harmony, but also glorious colors, pure-bred progeny, and breeding success.

In conclusion, I would like to refer to one other aspect involved in the successful care of these fishes. So far I have described to you the reactions of cichlids in particular situations. These behavior patterns are quite common. The behavior of individual cichlids, however, is influenced by yet another factor that can not be taken into consideration beforehand. This is the character of individual specimens. Innate characteristics,

Problems will only occur for the aquarist if certain individuals are particularly aggressive. There are cichlids that can be described as being deliberately nasty. Such animals can cause a great

Ophthalmo-tilapia nasuta, gold male.

such as aggression toward sibling species, are not always present to the same degree in various individuals. Consequently, cichlids can be more peaceful or more aggressive than usual with corresponding deviations in their behavior patterns.

deal of commotion and disarray in an aquarium. Generally, only the alpha animal (the strongest, most dominant animal) can take on this role. Since the behavior of such malicious specimens can rarely be changed by husbandry procedures, such a fish should be removed from the tank. The next strongest fish will then take its place. According to my frequent observations, older male cichlids make distinctly better alpha animals. Young, subadult males are particularly aggressive. This is probably because they want to (and they must) break into the established hierarchy. On the other hand, older fish are more inclined to opt for peace and quiet. On more than one occasion I have also observed that older males have attempted to separate fighting females, often with success.

Notice how intense the yellow is on this dominant male *Aulonocara baenschi.*

EXAMPLES OF HOW TO STOCK THE TANK

Although an optimal East African aquarium should have a volume of at least 240, better yet 450, liters and more, a few species of cichlids can also be cared for in smaller tanks. For instance, snail cichlids can be kept in tanks as small as 30 to 40 liters. A particularly popular favorite is *Neolamprologus ocellatus*. This is a rather cute little fish that grows to a maximum size of 6 cm. In its natural habitat it lives inside the empty shell of a snail of the genus *Neothauma* in Lake Tanganyika. In captivity, however, these cichlids will also eagerly accept empty garden snail shells, which are often similar to *Neothauma*.

To decorate such small tanks it is sufficient to use a few small rock structures for the back of the tank and a sandy bottom in the foreground. A few empty snail shells should be put on the sand. Any further "decorating" is done by the new tank inhabitants. Plants can also be added to the tank, since they will left alone by these small predatory fishes. The snail shells are buried in sand by the cichlids using two different techniques. The sand may be excavated from underneath the shells. Once the shell has slid into the depression, it is then covered over with sand. An alternate method is to

Neolamprologus ocellatus with its shell hideaway.

simply cover the shell with sand after it has been carefully positioned. The latter is done with the aid of the small pointed teeth and a clever twist of the body. A 30 to 40 liter tank can accommodate a single adult pair of *Neolamprologus ocellatus* or 4 to 5 juveniles of the same species. At least 2 snail shells should be available for each fish.

Other snail cichlids, such as *Neolamprologus multifasciatus,* are also recommended for beginning aquarists. This latter species lives in schools, so you should get a small group to start out with. For that reason you also need a pile of empty snail shells.

Some snail cichlid species are rarely ever available from aquarium shops. Yet the two mentioned above can easily be purchased. Moreover, *N. ocellatus* is only moderately aggressive and, indeed, quite cute. Breeding this species is not particularly difficult, so that there usually will be a few growing juveniles in the tank with the adults.

For somewhat larger tanks we can, apart from the already mentioned snail cichlids, also keep small *Julidochromis* such as *J. transcriptus, J. ornatus* and *J. dickfeldi.* We can also add *Chalinochromis brichardi* and *C. bifrenatus.* However, since both of these genera belong to the slender cichlids and consequently have similar requirements, only one of them should be included in the composition of the East African cichlid tank. On the other hand, keeping either of them together with snail cichlids should pose no problems. For instance, the cichlid community of a 50 to 80 liter tank could be made up of small groups of *Neolamprologus ocellatus* and *Julidochromis transcriptus,* or substitutions of other species of the indicated genera. *Julidochromis* and *Chalinochromis,* respectively, require rocky formations with narrow crevices. Pieces of slate are particularly useful for this purpose.

Tanks from 100 to 120 liters can have a larger community of cichlids. Apart from the already mentioned species, we can now also add the larger *Julidochromis marlieri* and the small mbuna *Labidochromis caeruleus.* But the latter comes from Lake Malawi! If you do not wish to keep cichlids from both Lakes Malawi and Tanganyika in the same tank, you may wish to do without that one and

replace it with another species. Unfortunately, there are only a few cichlid species available for this size aquarium. Most species definitely need more space.

For a Lake Tanganyika aquarium you can also keep, apart from small *Neolamprologus* species, *Altolamprologus calvus* and *A. compressiceps*. *Altolamprologus calvus* and *A. compressiceps* require small caves as hiding places. In addition, you can also keep some of the smaller *Xenotilapia* species for this size tank. Regrettably, these are rarely available. When keeping *Xenotilapia* species, you must avoid *Neolamprologus ocellatus* and other snail cichlids. Since *Xenotilapia* are primarily sand fish and constantly move along the bottom chewing (sifting) through the sandy substrate, they would always be attacked by snail cichlids.

Here is an example of a Lake Tanganyika tank of 100 to 120 liters volume: 3-4 snail cichlids (same species), 3-4 *Altolamprologus* (same species), 3-4 slender cichlids of the genera *Julidochromis* or *Chalinochromis* (select one species only). All these

cichlids are predators, so that apart from rocky structures, delicate plants can also be included in the tank decoration. For a Lake Malawi tank you should select smaller *Aulonocara* species and *Labidochromis caeruleus*. This sort of tank should be in excess of 200 liters, which could (for instance) accommodate 10 individuals each of an *Aulonocara* species and *Labidochromis caeruleus*.

This species is traded as *Aulonocara nyassa* 'red flash'. The photograph shows a subadult that has not yet reached full adult coloration.

Labidochromis caeruleus is bright golden yellow (at least one morph is) and the dorsal fin has a black submarginal band. It grows to a maximum length of about 9 cm and presents an attractive contrast to the mostly blue *Aulonocara* species, which are fully grown at 10 to 12 cm length. These species need a lot of rock formations in the tank. Both belong to the polygamous mouthbrooders.

Tanks in excess of 240 liters can be stocked with some of the smaller *"Haplochromis"* species, as well as species of *Pseudotropheus*. An aquarium of that size could be stocked as follows: a group of *Aulonocara* (5 to 6 individuals, all of the same species), 7-8 *Labidochromis caeruleus*, and one pair of *"Haplochromis."* An example of a Lake

Tanganyika tank: one pair of snail cichlids, a group of 5-7 *Callochromis pleurospilus* (or alternatively *Xenotilapia*), 5-6 smaller slender cichlids (*Julidochromis* or *Chalinochromis*), and 6-7 individuals of the small species of the genus *Ophthalmotilapia*, (e.g. *O. ventralis*).

The next tank size, of about 400 liters or more, now affords us the opportunity to obtain a diverse selection of available cichlid species. Tanks of this size can be used to care for cichlids from Lake Malawi and Lake Tanganyika, even in larger groups. Generally, however, aquarists tend to prefer greater species diversity in favor of keeping large groups (and fewer species) of cichlids. Yet, some species should always be kept in suitably large numbers.

If there is any intention at all of eventually breeding particular species, the possibility of crossbreeding or hybridization should always be kept in mind when selecting particular species. Otherwise, newly hatched young may be eaten by the other aquarium inhabitants, so that hybrids possibly will not survive. Therefore, it is better to avoid hybridization right from the start. Genetic crosses can be avoided as follows: select only one *Aulonocara* species of one color variety (the former *Trematocranus* species are to be included in this), and in "*Haplochromis*" species watch for the spotted or striped pattern in females and the coloration of the male. The predominant forms in Lake Tanganyika are *Tropheus* species, which are represented in numerous color varieties. Since they crossbreed to a limited extent only, you will find an appropriate table below, which contains the most commonly found species.

The following *Tropheus* species and varieties can be kept together without risk of interbreeding on the condition, though, that both sexes are present from each species or variety. Otherwise, individual specimens will crossbreed with other varieties.

The popular *Tropheus duboisi* also has four color variants. Therefore, when you acquire cichlids, make sure you know exactly what species or subspecies you are dealing with. Color varieties frequently also have a location designation added to their specific name, e.g., *Tropheus moorii* Katanga. *Tropheus moorii* is

Another slender cichlid, *Chalinochromis bifrenatus,* is readily available in the trade. It is considered to be a very peaceful slender cichlid. This species deposits small greenish eggs inside crevices.

This graph indicates which *Tropheus* species and varieties can be successfully maintained together. If one does keep more than one species or variety together, then at least one pair of each should be provided so as not to encourage possible hybridization.

X=Compatible species and varieties

	Duboisi	Katonga	Bulu Point	Chimba	Moliro	Chaitika	Polli	Ikola	Bemba	Chipimbi	Nangu	Red Rainbow
Duboisi		X	X	X	X	X	X	X	X	X	X	X
Katonga	X		X	X	X	X				X	X	X
Bulu Point	X	X		X	X	X	X			X	X	X
Chimba	X	X	X				X	X	X			
Moliro	X	X	X				X	X	X			
Chaitika	X	X	X				X	X	X			
Polli	X		X	X	X	X		X	X	X	X	X
Ikola	X			X	X	X	X			X	X	X
Bemba	X			X	X	X	X			X	X	X
Chipimbi	X	X	X				X	X	X			
Nangu	X	X	X				X	X	X			
Red Rainbow	X	X	X				X	X	X			

the scientific (Latin) name and 'Katanga' designates the location where this variety was originally found in Lake Tanganyika. Since not all varieties of a particular species are always available, it is advisable to acquire a large group, especially of rare species. Juvenile specimens and subadults often do not show their full adult coloration, which often makes an identification impossible. As responsible aquarists you should also watch for minor (coloration) details in order to avoid hybridization.

At this point it is certainly not possible to describe in detail all suitable cichlids occurring in the East African lakes. Still, I would like to mention a few of the species. Relatively peaceful are *Xenotilapia* species as well as *Cyprichromis* species (both from Lake Tanganyika). However, in an aquarium, species of both genera are initially rather timid. Their only defense is a hasty flight, and, particularly in a newly set up tank, they tend to go easily into a panic when frightened. They will settle down, though, once they have adapted to their new surroundings. Both genera must be kept in schools,

which conforms to their natural behavior in the wild.

Callochromis species are sometimes extremely aggressive, especially *C. macrops*. They require tanks from 450 liters upwards. Their dominance is astounding. Even aggressive grazers, such as *Tropheus, Petrochromis,* and *Simochromis* can be intimidated by these fish. They can be kept together with larger *Julidochromis* species, *Neolamprologus tretocephalus*, and *N. sexfasciatus*. On the other hand, *Callochromis pleurospilus* is much more peaceful and therefore quite suitable for a habitat tank together with other species.

Labidochromis caeruleus from Lion's Cove, Lake Malawi, is one of the more peaceful mbuna available in the hobby.

This *Tropheus brichardi* from Katonga, Tanzania, is one of the most aggressive members of the *Tropheus* complex to date.

Small to medium-size *Neolamprologus* species such as *N. leleupi, N. pulcher, N. savoryi, N. brichardi, N. ocellatus*, and others can be kept in large aquariums without problems. Large *Neolamprologus* species, such as *N. tretocephalus* and *N. tetracanthus*, are predators that are extremely aggressive as adults.

In large aquariums, *Tropheus* species can easily be kept in larger groups, but they are not suitable to be kept together with a diverse tank population of peaceful cichlids. They are very active and domineering fish that tend to intimidate and suppress other species. Aquarists who like to keep *Tropheus* species are best advised to set up a tank for keeping only *Tropheus*. Anyone with a habitat tank who insists on *Tropheus* should confine themselves to a single pair.

Comic relief in an aquarium can be provided by cichlids of the genera *Tanganicodus, Eretmodus*, and *Spathodus*. These bottom dwelling fishes have somewhat subdued coloration but are characterized by fascinating behavior. Because of a poorly developed swim bladder, the swimming behavior of these fishes looks more like "hopping" rather than outright swimming. Regrettably, however, these cute little fishes are rarely ever available in the aquarium trade. Reasons for this seem to be that they are rather delicate to transport and they are susceptible to certain diseases.

Petrochromis and *Simochromis* are extremely aggressive. These mouthbrooding grazers from Lake Tanganyika should only be kept after quite some experience has been gained with other East African cichlids. For the care of these fishes I also recommend a tank of at least 500 liters volume and a large group of at least 15 to 20 fish. Only this will minimize the aggression among them, but even then substantial injuries and even mortalities can occur. Hiding places and a sufficient number of males lessens the aggression among males and towards the females.

Particularly recommendable for a Lake Tanganyika tank are the genera *Cyathopharynx* and *Ophthalmotilapia*. The latter genus is particularly suitable for maintenance in home aquaria. *Ophthalmotilapia ventralis* grows to a maximum size of about 12 cm. Since this species will spawn on flat, smooth rocks as well as in self-excavated sand pits, they are suited principally for tanks in the size range of 250 to 500 liters. For this species we should also attempt group care. In my experience a single male in a group will only rarely display magnificent coloration (usually only during courtship), so it is advisable to have at least 2 males in a group of, say 7 fish. The continuous presence of a rival definitely promotes a more permanent, intensive coloration. Other advantages to keeping this cichlid are its moderate aggression and easy maintenance in an aquarium of this size range. There is rarely any mouth-tugging among rivals, and when it does happen it occurs only very briefly. I have never

Pseudo-tropheus sp. "*acei.*"

observed any fatal injuries. Large fishes, such as *Cyathopharynx furcifer*, require tanks from about 450 liters upwards in order to allow them to dig large sand nests.

Some of the best known cichlid genera from Lake Malawi include *Pseudotropheus, Aulonocara,* and "*Haplochromis.*" Species of the genus *Pseudotropheus* should be kept in tanks of at least 350 liters volume. In order to keep these fishes successfully they must be kept in groups. They can also be kept together with other cichlids from the rocky or rubble littoral zone. This reduces intraspecific aggression.

On the other hand, *Aulonocara* can easily be kept in pairs. Even small groups of 3 to 7 fish will do well together. For small groups it is, however, advisable to include only 1 or 2 males; in larger tanks (450 liters and greater) we can add more males. Mouth tugging and other destructive fighting is rare. The defeated male will often lose its magnificent coloration; it will take on the drab, inconspicuous color pattern of females and even swim with the female group. Species of *Aulonocara* are suited for community tanks

containing small species of "*Haplochromis.*" These can include "*Haplochromis*" *fenestratus*, "*H.*" *chrysonotus* or "*H.*" *borleyi*. These "*Haplochromis*" species are mainly suited for smaller tanks, i.e., from 240 to 400 liters.

Large tanks can be stocked mainly with large "*Haplochromis.*" Some of the more peaceful of these are "*Haplochromis*" *moorii*, "*H*". *electra*, "*H*". *euchilus*, and "*H*". *quadrimaculatus*. These reach sizes of about 15 to 25 cm, but still remain peaceful toward smaller tank inhabitants.

Rather peaceful mbuna are the *Labidochromis* species. *L. freibergi, L. caeruleus,* and *L. pallidus* are smaller species with maximum sizes of 7 to 10 cm. Because of their moderate aggression and small size, the species can be kept in small tanks. On the other hand, *Labeotropheus* species should only be maintained within groups kept in larger tanks. Their intraspecific aggression is well defined. Even females will generally establish territories. They can be kept together with moderately active "*Haplochromis*" species.

Other small size (about 10 cm) mbuna are the *Cynotilapia* species. They, too, can be kept in tanks from 240 liters upward, together with other mbuna. Their behavior is largely similar to that described for *Pseudotropheus*. Also exhibiting strong intraspecific aggression but peaceful toward other aquarium inhabitants are the *Petrotilapia* species. Because of their size—15 cm and more, depending upon the species—they can be kept in tanks of 450 liters and larger together with other large mbuna.

Species of *Melanochromis* also live in the rocky littoral zone. Intraspecifically, these medium size mbuna are considered to be very aggressive. Even keeping a single male with a group of females is often critical. Generally more successful, however, is keeping a larger group of, say, 15 to 20 fish. This leads to a more peaceful situation in a species tank as well as together with other mbuna. It is pointless to acquire only a few fish since there will be immediate fighting with inevitable fatalities. This aggressive instinct must always be taken into account when planning to keep these fish.

The numerous species of the genus *Lethrinops* from Lake Malawi can be considered as sand fishes. These cichlids are not widely kept in aquarium circles. Their dietary requirements are similar to the cichlids of South America. They "sift" edible materials from the bottom substrate. *Lethrinops* species are not very domineering; they can be

Xenotilapia ochrogenys is fairly peaceful but frightens easily.

kept together with or in conjunction with some of the more peaceful "*Haplochromis*" species. The establishment of territories among the various males can be controlled by separating individual sandy patches with rocks or plant thickets. *Lethrinops* species range in size from 10 to 20 cm (depending upon the species) and should be kept in medium to large size tanks.

PURCHASING TIPS

In order to avoid buying the wrong fishes you should have it clear in your mind how you want to set up and then stock your tank. Detailed descriptions of the appropriate species can be found in the relevant literature (aquarium books, journals, etc.). Advice from experienced aquarists should also be solicited, particularly if they are known to have kept some of the species you are considering. Breeders can also be a virtually inexhaustible source of information, since they may have kept certain cichlid species for many years. Therefore, you are advised to thoroughly research all aspects of the cichlid species to be kept before the purchase. This may save you a lot of trouble and disappointment later on. When selecting from among the appropriate cichlid species, keep in mind that not all cichlids desired are actually available. So, develop a "fallback" position by selecting certain substitute species. Since it is often difficult to remember scientific (Latin) names, you should make appropriate written notes.

All cichlids occupy specific ecological niches and this factor must be taken into consideration when making up your "shopping list". Unfortunately, specific advice about cichlids is usually only available in specialist aquarium shops. Again, the special reference literature and a detailed shopping list can overcome this hurdle.

When purchasing fishes you must concentrate on the general condition of the fishes for sale. Those that look "beat up", sick, or appear in poor overall condition (thin!) must not even be considered. In such cases do without it, even if it should be a very desirable species or one you have been looking for for a long time. Proper signs of a healthy fish are normal activity (swimming behavior) and active

feeding. Especially the latter can divulge a lot about the condition of the fish. If you have an opportunity to watch it feeding, take advantage of it. Develop a discerning, critical eye! If a particular cichlid species is available from several sources, compare the condition and behavior of the fish at the various potential suppliers. Rely more on your own judgment than obvious "quick sale" attempts, such as "it will become well again," or "it was eating just yesterday." A purchase of clearly sick fishes carries incalculable risks! Simple diseases, such as ichthyophthiriasis (White Spot), which you think you can cure with a few drops of medication, are often

Cyprichromis leptosoma is very peaceful. It does best in a larger group.

only secondary symptoms. In actual practice, that means that you are really only treating a minor pathological problem and NOT the principal disease. It is often difficult to localize the main disease, which may even be incurable. In that case proper treatment may be impossible, even for those aquarists experienced in the treatment of fish diseases. Should you still insist on buying the fish, it must undergo a proper quarantine period in a separate tank and with the appropriate simultaneous treatment.

Apart from inspecting obviously sick fishes you must also take a close look at those specimens that appear thin and are lethargic. These can also be potential disease carriers or the symptoms can be due to a nutritionally deficient diet. A deficient diet can lead (minimally) to abnormal (retarded) growth and it often causes diseases. Not all cichlids are alike. If you are uncertain about the condition of a particular specimen, keep away from it or seek the advice of an experienced aquarist. A cursory review of fishes available from different sources inevitably reveals variable qualities and different prices. The latter should be of subordinate importance. With East African cichlids we must also keep in mind the fact that with maximum life spans of 10 to 15 years these fish can get rather old.

Those species that have been kept for a long time and that are properly cared for tend to show rather spectacular colors and considerable trust toward their keeper. After a few years large cichlids tend to become almost hand-tame, a sign of confidence. For species that are rarely available in the trade it may be advisable to acquire one or two extra specimens. This is particularly important in those cases where group harmony is dependent upon numbers of individuals in the group. Maturity in cichlids is frequently not reached until after 2 or 3 years, but two years is a long time span in an aquarium situation. During these two years a fish may manage to jump out of the tank through an inadvertent opening in the tank cover – some may die of diseases or as a consequence of injuries sustained in severe fighting. While these are exceptional circumstances, they should certainly be taken into consideration when stocking an aquarium. It may be difficult later on to replace these specimens with others of the same species, color variety, size, or sex. Moreover, adding another specimen to an established fishes can be sick or very aggressive or unwilling or unable to mate satisfactorily. Although this is certainly not the rule it is, nevertheless, a factor to be kept in mind. A well-bonded breeding pair or breeding groups are rarely

group or community has its inherent problems. Therefore, when buying rare cichlids, get an extra one or two...if you can afford it!

"Bargains" can indeed be good opportunities. However, they may not always be as opportune for the buyer as they may be for the seller. In actual practice this means that experienced aquarists sometimes exploit the inexperience of others in order to get rid of undesirable specimens. The ever available at "dumping" prices, since the seller will always be aware of their proper value. This also applies to other specimens that are healthy and in supreme condition. Therefore, it is important to remain alert and to keep the above-mentioned points always in mind when buying cichlids. So-called opportunities must always be thoroughly examined before considering purchasing "bargain" specimens.

Finally, I would like to

The graceful appearance is deceiving. *Callochromis macrops* may be a very attractive fish but it is not easy to keep.

discuss the various water quality conditions. There are substantial regional difference in water quality parameters, but this becomes significant only when the water in the natural habitat of the cichlid is distinctly less alkaline. I have often heard other breeders say that transferring these cichlids from very weakly alkaline water to strongly alkaline water presents problems. The effect occurs only after a few weeks in the new tank and manifests itself with or without disease symptoms, but inevitably leads to the demise of the individuals involved. Presumedly, these cichlids are capable of adapting to alkaline water, especially through successive generations as captive bred stock. On the other hand, fish like that appear to have great difficulties adjusting after a sudden transfer to water of endemic (natural habitat) conditions. The reason for this inability to adjust has yet to be fully researched.

CARE AND MAINTENANCE

HOW MANY FISH PER TANK?

Apart from the selection criteria already discussed for purchasing cichlids, it is also of importance to determine how many fishes can be placed into a tank of a certain size. The basic calculations for establishing the correct number of cichlids per aquarium are based on size and volume of the tank. The size of the cichlids refers to the final adult size. Experience has shown that the maximum sizes given in the aquarium literature are often exceeded when the fishes are kept under optimal aquarium conditions. Optimal care in the form of diet, filtration, water changes, and adequate space all contribute to achieving a maximum size. Therefore, the listed sizes for various cichlids are best considered to be minimum sizes.

The basic calculation for stocking a tank is: Each centimeter of fish length requires 1.5 to 2 liters of water. If a tank is decorated with a lot of rocks the volume unit/cm fish length should be increased. For example: If you have a tank with the dimensions 150 cm X 60 cm X 50 cm, this gives a water volume of 450 liters. This tank is decorated with rocks, but not excessively so. The aquarium volume permits the maintenance of (450 liters / 2 =) 225 cm of cumulative fish length. If all future specimens designated for that tank grow to a maximum size of 12 cm, the tank could be stocked with (225 cm / 12 =) 19 specimens. If the tank is relatively sparsely decorated the total aquarium population could be exceeded by a few specimens. A correct stocking rate facilitates optimal care and reduces maintenance work.

WATER QUALITY MAINTENANCE AND WATER CHANGES

According to various authors the following water quality parameters are endemic for the natural habitat of Lake Malawi: pH 7.7-8.8; total hardness 3-6; carbonate hardness 5-8;

temperature from 23 - 28°C. In the upper water layers (where most of the Lake Malawi cichlids live) the temperature is relatively constant at 26-27°C. The indicated values for pH, total hardness, and carbonate hardness can be exceeded without creating any apparent problems. On the other hand, softer water can be problematic. The tolerance for softer water is greater in Lake Malawi cichlids than among Lake Tanganyika cichlids. Lake Malawi cichlids are frequently being kept at a pH slightly above pH 7. This, however, is not an ideal situation and should only be considered to be an exception. The well-being of these cichlids depends clearly on the correct water quality parameters, the most important of which are the pH and temperature.

For Lake Tanganyika the following values have been recorded: pH 7.5-9.3; carbonate hardness 15-18 degrees; total hardness 10-12 degrees. The temperature was relatively constant at about 26°C. Even at greater depths the temperature drops by only 3°C. Lake Tanganyika cichlids can also tolerate distinctly higher pH, total hardness and carbonate

Labeotropheus fuelleborni displays strong intraspecific aggression. Group maintenance is advantageous.

hardness values. The low end of the ranges indicated can be considered as the minimum for both lakes. For Lake Tanganyika cichlids it is important to make sure that the pH values do NOT drop below the indicated range. Aquarium experiences have shown that there can be little tolerance outside the listed range.

As already mentioned, higher alkaline values do not present any problems for these cichlids. On the other hand, softer water must be treated to increase its hardness components. If anybody's municipal water supply deviates substantially from these values he or she might wish to reconsider setting up an East African cichlid tank.

If a frequent large water change is to be anticipated, conditioning this water may only be possible with considerable logistic effort and at some expense. If your municipal water supply is marginal with regard to the above water quality parameters and you are going ahead with keeping East African cichlids, there are a few points you must keep in mind. Since tap water can vary in quality it is important to constantly monitor the above parameters. You must also take into consideration that in a newly set up tank the water can change due to influences caused by the substrate and any rocks that may have been used for decoration. Also, in order to avoid a further drop in the pH value, you must NOT fertilize the plants with CO_2.

On the other hand, there is often a decline in water hardness and pH in tanks that have been operating over a long period of time. The changes referred to always apply to the original tap water. A slight increase in hardness can be achieved by using calcareous rocks as decoration. You can also use coral sand (coral rubble, shell grit) as the filter medium (e.g. in biological trickle filters). Special salts commercially available for Lake Malawi and Lake Tanganyika cichlids also increase water hardness slightly. If your tap water shows water quality values that are distinctly below the desired range, you have to initiate massive corrective measures. If the pH value is too low (which is rarely the case), you can elevate it with the appropriate chemicals (commercially available). Generally, though, it is the pH AND the water hardness that are too low. Only chemicals can then correct this situation. By cautiously adding diluted sodium hydroxide (lye) solution, the pH and carbonate hardness can be increased. The total hardness can be elevated by adding a solution of calcium sulfate. Both of these preparations are available from drug stores. It is advisable to do some tests first on a tank without fishes, since every water type requires different adjustments. The preparations and their effects can also be tested in a bucket of water. All

Salts identical to the salt found in Lakes Malawi and Tanganyika are available at your local pet shop.

amounts added as well as the water values should be meticulously recorded. By necessity one should be careful to slowly adjust the pH so as not to cause pH shock among the aquarium residents.

If there is to be a water change in a fully stocked tank, the make-up water will have to be adjusted before or while the tank is being refilled. It is often easier to put all make-up water in a separate container first and then add the required chemicals before the water is added to the tank. In this way, only diluted solutions will get directly into the tank.

In any event, it is imperative that all water parameters are closely monitored, especially in the critical areas. Accordingly, the temperature must be frequently checked. The correct habitat temperature must be maintained. For East African cichlids the required temperature range is 26° to 28°C. For juveniles it is best to aim for the upper end of this range,which enhances their growth. A correct

Ophthalmo-tilapia ventralis is an attractive fish that is mildly aggressive. With a maximum size of about 12 cm, it is also well-suited for medium size tanks.

temperature is also important for the well-being of the fishes. The warmer the water, the more active the fishes are. Moreover, increasing the temperature by 1 or 2 degrees promotes courtship activities and eventual breeding.

A frequently asked question is how often to change water. There is no generally valid answer for this. The need to change water is influenced by many factors: number and size of fishes in the tank, tank size, filtration, and volume of make-up water all play a role. The better the water conditions are, the less frequently water has to be (partially) exchanged. Nevertheless, a few fundamental guidelines can be provided.

Due to the excretion of urine by the fishes, the tank water will become initially yellowish and later on it will turn solid yellow. The clearer (whiter) the water is, the smaller the amount of urine in the water. East African cichlids, which must have crystal clear water, should not be permitted to swim in yellowish, cloudy water. Consequently, the water must be changed when, if not before, it has a yellowish tinge. It is, however, imperative that maximally only 50% of the

tank volume is to be replaced with make-up water. At the same time it is important to neutralize the make-up water before adding it to the aquarium, the purpose being to avoid shocking the tank inhabitants.

Suitable test kits are also available commercially for determining nitrite levels and concentrations of other harmful substances. Small,

"Haplochromis"
quadrimacu-
latus. A very
attractive
representative
of a
haplochromine,
but one that
is moderately
aggressive.

frequent water changes are better than massive water changes over long time intervals. For large tanks this may involve a considerable volume of water, and so it may be advisable to use a garden hose. When equipped with proper fittings, such a hose is easily and quickly attached to any household tap. Suction cups can be used to secure the other end of the hose to the inside of the tank. This avoids water being splashed around. It is useful to turn water changes into a regular routine, so that they cannot be forgotten. Smaller water changes are unavoidable when the tank has first been set up. Since it takes many weeks for a tank to become fully broken in, any substantial

There are products available for the instant conditioning of aquarium water.

Facing page: Python makes a water changer that makes the chore of changing water more pleasant.

binding the harmful substances. If the tap water is heavily chlorinated, it is advisable to spray it against the inside of the tank. The splashing action will dissipate a large part of the chlorine.

water change would inhibit the development of nitrite reducing bacteria.

There are a number of solutions to the various modern-day water problems. Commercial sources (aquarium and pet shops) offer water treatment agents that provide comprehensive protection against chlorine, neutralization of all dissolved salts, and binding of all harmful metal ions. At the same time the mucous layer over the skin of fishes is fortified and stabilized. Sometimes there are also problems with the presence of insecticides and pesticides in the tap water. This is particularly relevant in agricultural areas. Activated charcoal provides the answer for problems like that by

For cichlids, the above-mentioned aids are normally not of vital consequence, except under some of the extraordinary circumstances described above. In regions where the tap water has a reduced iodine content, Malawi-Tanganyika salt can be added to the tank. Iodine deficiency can cause the formation of tumors below the gill covers in some cichlid species. Cichlids particularly prone to develop this condition are species of *Julidochromis*, *Lamprologus*, *Telmatochromis*, *Chalinochromis*, and *Xenotilapia*, as well as *Altolamprologus calvus*, *A. compressiceps*, *Neolamprologus ornatipinnis* and *N. leleupi*. The addition of iodine or the above-mentioned special Malawi-Tanganyika salt can eliminate this condition and even produce healing in affected fishes.

Newly purchased cichlids, like all other new fish acquisitions, must be

CLEAN and FILL

NO SPILL

AQUARIUM MAINTENANCE SYSTEM

THE AQUARIUM OWNER'S FRIEND

PYTHON
PRODUCTS, INC.

422 N. 15th Street
Milwaukee, Wisconsin 53233
(414) 273-1313
FAX (414) 273-2213

quarantined. Unfortunately, there is often a mistaken belief that this is not necessary. It can really only be avoided when a tank is newly set up and does not contain any fishes. This, in effect, becomes the quarantine tank. Generally, any transfer of fishes can cause an outbreak of disease. Stress due to transport and the anxieties involved can lead to a state of discomfort, the precursor of diseases. Therefore, healthy fishes can become sick after they have been transported. The problem (of stress-related diseases) is further compounded by the changeover into water with different water quality parameters. In any event, caution must be exercised. After all, the fishes already in the tank must not become infected with disease carrying organisms from newly arrived specimens. The quarantine period should last for at least 2 weeks, preferably for 3 to 4 weeks. Different species of fishes are not equally susceptible to the various diseases. Therefore, it can take a variable amount of time for disease outbreaks to occur. The size of the quarantine tank depends upon the fishes to be kept in it. Yet, even small species and juvenile

fishes should be given at least a 50 liter tank. Large specimens need a tank of 80 cm to 100 cm length. A quarantine tank does not have to be permanently set up, only as required. Water

Neolampro-logus tretocephalus male guarding his brood territory.

for the quarantine tank can be taken from larger, well-established tanks. Similarly, utilizing previously used filter media facilitates a rapid establishment of a quarantine tank.

Newly arrived specimens have to be closely monitored during the quarantine period. Any feeding must (at first) be done cautiously and sparingly. This is particularly important for wild-caught fishes that, after a long, arduous period in transit, must be acclimated gradually to new conditions and a different diet. A prophylactic treatment with medication is not necessary. This should only be done once there is a definite disease outbreak, otherwise the fishes could become resistant to medication. Should there be another outbreak of the same disease later on, it could well be possible that the same medication will no longer be effective at the same dose levels.

Special problems can arise when cichlids of different origins share the same aquarium. This applies to the main tank as well as to the quarantine tank. Under these conditions various diseases can occur more or less simultaneously, which are then cross-transmitted among the newly arrived specimens. This, however, is an exception.

Nevertheless, a definite quarantine period cannot be avoided. Introducing sick fishes into an established aquarium can cause much damage. Do not expose your own fishes to that sort of risk. For ease of maintenance and to assure quick acclimatization of the fishes under quarantine, the tank decor should be kept simple. Hiding places can be provided in the form of earthenware flower pots, maybe a few rocks, and some other cover. If several tanks are in use, the quarantine tanks should be close to the other tanks for visual purposes. Visual contact between fishes contributes to their well being and they will lose their shyness much faster. Fishes that are frightened behave abnormally. These fishes sometimes do not even come out of hiding to take food. This then makes it difficult to monitor the fishes for diseases. Once the quarantine period (and possibly also any treatment) has been completed, the newly acquired fishes can safely be introduced into the main tank.

INTRODUCING NEW FISHES

Problems can occur when new fishes are introduced into your already established tank. These are much easier to handle when you are dealing with juveniles than with adults. Since a group of young fishes can be put into the tank simultaneously, all available space can be split up correspondingly among the fishes, even though young cichlids are usually also somewhat territorial. Such minor aggression has rarely ever any serious consequences. Territorial fights among subadult cichlids usually do not involve injuries among the rivals. At that stage—shortly after introducing all juveniles—the available tank space is quickly divided up among all fishes. Cichlids that grow up together usually behave less aggressively toward each other as adults. The reasons for this have already been discussed.

It is much more difficult to add new specimens to an already established cichlid community. First, we must keep in mind that small fishes could be considered as food by some of the adult fishes in the tank. Prior to completing the purchase you must make sure that the individuals under consideration will fit into the tank as a species and in the available size category. Particular attention must be paid to

These fish like to feed on 'scalded' lettuce or spinach. *Tropheus duboisi* (Kigoma) is considered to be the most peaceful of the *Tropheus* species.

those species already in the tank that could be potential predators.

The size difference between fishes to be kept together must not be more than a few centimeters. Cichlids less than 5 cm long must not be placed in a large aquarium if there are already a few large individuals occupying it. Under such circumstances juveniles have difficulty feeding. Most are very shy and frightened. If they are placed under too much stress, it can well lead to disease problems. Leaving a hiding place in search of food is a great risk for young fishes. The more aggressively feeding larger fishes can injure or possibly even kill some of the young fishes. Because they are suppressed under these circumstances, the young are not getting optimum care, which then leads to retarded growth. Moreover, large fishes are usually fed with larger food items, and so from that point of view there is an insufficient food supply for young fishes. Therefore, it is advisable to rear juvenile fishes in a separate tank until they have reached the proper size.

Introducing new fishes into a tank should not be done at the same location where, normally, food is given. Tank inhabitants waiting to be fed will rush at the anticipated food. Frequently the resident fishes will bite the new arrivals before realizing that the new fishes added are not food. By the time this is realized, the newcomers may have already been severely bitten.

Adding adult cichlids to an established cichlid community requires a few "tricks" in order to achieve peaceful assimilation. It is advantageous if the tank is redecorated before the new fishes are added. This virtually destroys existing territories so that there is room for a new hierarchy and new territories. If this is unsuccessful, the tank can be maintained for a few days totally without any decorations but retaining the existing fish community. This way established residents and the newcomers can get used to each other. Under such circumstances one would not expect severe aggression since there are no territories to be maintained. After a few days everything is put back into the tank but arranged differently, so as to avoid the reestablishment of old territories. Instead, additional territories should be included for the newcomers.

If the aquarium decorations cannot be removed, attempts should still be made to change the established territories. Observing your fishes will quickly show you that there are invisible territorial borders. These are generally indicated by certain "landmarks," such as particular plants, plant this territory simply by moving the plant. This little trick usually goes unnoticed by the territorial occupant. This way existing territories can be made smaller, in order to create new ones.

It is usually rather difficult to add more specimens of a particular species that is already being kept in that tank.

thickets, rocks, caves, and items of equipment inside the tank. Territorial cichlids patrol their particular territory regularly. You can easily determine the size of the territory from the extent of such patrols. For instance, if you observe that a certain fish has its territory up to a large center plant, you can adjust the size of The same applies when a suitable breeding partner is subsequently added. In both cases we proceed on the premise that the tank occupants have a distinct "home" advantage. For that reason it is always better that any subsequent additions to a tank are purchased slightly larger than those in the tank. To some degree this nullifies

Melanochromis species are best kept in groups. The males of *M. auratus* have bluish background colors, the females yellowish colors.

Tropheus duboisi male. The literature gives a maximum size of 14 cm. Aquarium specimens tend to get larger, as seen in this 5-year-old captive-bred specimen.

the home advantage. If a male of a species is added to one or more females, there are rarely ever any problems. The new male will quickly become dominant and is recognized by the females as such. If another male is to be placed into the same tank, the dominant male can be removed for a few days. This gives the newcomer the opportunity to become acclimated. After two to three days the old male can then be returned. It may be necessary to repeat this exercise a few times until both males are properly established. If one male persistently pursues one or more of the newly added individuals, the aggressive, dominant male should be removed. Providing additional hiding places (caves) or sight barriers usually does not bring the desired success. If the newcomers are found in their hiding places, the ensuing chase is that much more intense.

Hierarchies tend to change or are completely abandoned during darkness. Consequently, darkening a tank can provide an advantage for newly introduced fishes. If this is done during the day, once the new fishes have been added the tank is darkened. Alternatively, introducing

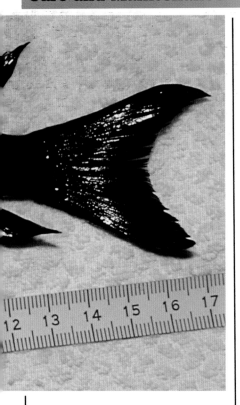

new fishes can be done just before it normally gets dark. Phases of subdued lighting and darkness can also enable newly added fishes to become more quickly and better adapted to a new environment. Most newly added fishes have high respiratory rates. In that condition they have no opportunity to resist the established tank occupants. They are not (yet) aware of the spatial situation, the tank decoration, or the territories of individual fish. Providing a darkened tank condition will help to resolve this problem.

Sometimes there is no time to separate rival fishes, inside or outside the tank. For example, you have an important appointment. What to do now? Easy, send your fishes to bed! Turn the lights off. If there is still a lot of ambient light reaching the tank, an additional blanket is placed over the tank. This inhibits any further activities. In your absence the fishes will not harm each other. This measure must only be considered as an "emergency action." Under no circumstances must the tank be kept in darkness longer than necessary.

The assimilation of new fishes can be simplified if a large number are introduced simultaneously. The more newcomers that are added to the tank, the simpler acclimatization becomes. The reason for this is the fact that well-established tank inhabitants can no longer concentrate on individual specimens from among the newcomers.

None of the preceding methods for subsequent acclimatization is guaranteed to succeed. A combination of as many different methods as possible is often the quickest path to success.

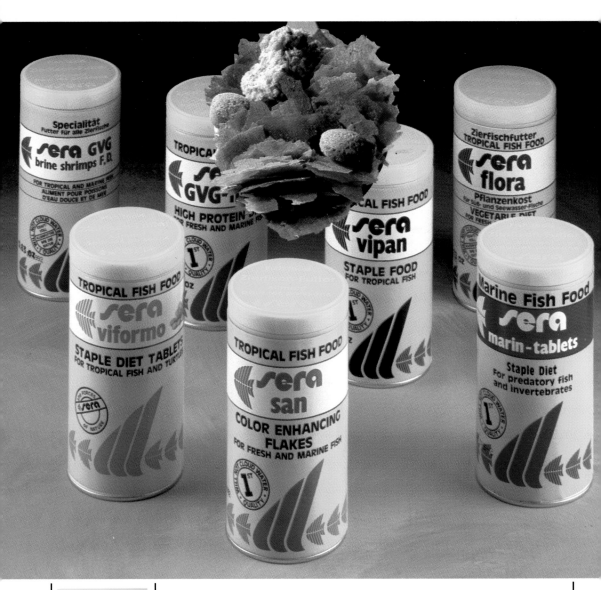

Your local pet shop should have a wide variety of flake foods, pellet foods and frozen foods. Feed variety, not just the same food all the time.

FOOD REQUIREMENTS

A very important and complex factor in tropical fishkeeping—in our case keeping cichlids—is the food for these fishes. First, we must raise the question as to the composition of the fish food. Tropical fishes can roughly be divided into plant-feeders (herbivores), grazers (limnivores), those feeding on different types of food (omnivores), and meat eaters (carnivores). An experienced aquarist, aquarium shopkeeper, or the relevant literature will tell you into which category your fishes belong.

The following listing of vitamins and their functions provides information about the source, effect, and deficiency symptoms of the essential vitamins. NOTES: Vitamins A and E are only effective if given at the same time. The listed vitamins, minerals, and trace elements are already present in commonly used dried foods and have been accounted for. The addition of liquid vitamins is advantageous, particularly since some manufacturing procedures reduce or destroy vitamins. In addition, prolonged storage also destroys vitamins. The situation is different with minerals and trace elements; they are not affected by long storage periods. Vitamin supplements are essential if fishes take little or no dried food at all. It is common for newly wild-caught fishes to take dried food only very reluctantly.

Vitamin A (water soluble)
Effect: Cell growth, especially in juvenile fishes

FOOD REQUIREMENTS OF CICHLIDS

Fish Category	Protein	Fat	Bulkage	Raw fiber
Herbivores	12-30%	1-3%	54-70%	5 -10%
Limnivores	30-40%	2-4%	39-45%	2 - 6%
Omnivores	32-40%	2-5%	36-52%	3 - 8%
Carnivores	45-75%	3-5%	5-39%	2 - 4%

Deficiency symptoms: Poor growth, deformation of vertebral column and fins

Source: Liver, cod liver oil, carrots, spinach, dairy products

Vitamin B₁ (water soluble)

Effect: Breakdown of carbohydrates, supports nervous system, promotes digestion, growth and fertility

Deficiency symptoms: Poor growth, loss of appetite, frightened behavior

Source: Peas, carrots, spinach, paprika

Vitamin B₂ (water soluble)

Effect: Protein structure, control of enzymes

Deficiency symptoms: Poor growth, reduced appetite, and clouding of eyes.

Source: Spinach, peas, carrots, paprika

Vitamin B₃ (water soluble)

Effect: Breakdown of food for endogenous protein

Deficiency symptoms: Weakness, poor digestion, aimless movements, later on development of tumors possible

Source: Peas, spinach, liver, paprika

Vitamin B₅ (water soluble)

Effect: Control of hormone production and metabolism

Deficiency symptoms: Cell degeneration, adhering gill membranes, general weakness

Source: Paprika, liver

Vitamins B₆ & B₁₂ (water soluble)

Effect: For enzyme system, protein metabolism

Deficiency symptoms: Increased respiration, loss of appetite, stunted growth and shyness.

Source: Liver, paprika

Vitamin C (water soluble)

Effect: Tooth and bone formation, improved healing, digestion, formation of cartilage.

Deficiency symptoms: Changes in skin, liver, kidney, and muscle tissue

Source: Paprika (red), spinach, peas

Vitamin D₃ (fat soluble)

Effect: Bone structure

Deficiency symptoms: Degeneration of bone structures

Source: Fish liver, fish meal

Vitamin E (fat soluble)

Effect: Development of sex organs; fertility

Deficiency symptoms: Infertility

Source: Eggs, cereal, pasta (noodles, etc.)

Vitamin H

Effect: Cell growth

Deficiency symptoms: Insufficient blood formation

Source: unknown

Vitamin K (fat soluble)

Effect: Blood formation,

blood clotting
Deficiency symptoms:
Death following injuries
Source: Lettuce, peas,
spinach, chard

Vitamin M (water soluble)
Effect: Blood formation and
metabolism, control of
sugar
metabolism
Deficiency symptoms: Dark
skin pigmentation,
weakness, changes
in liver, kidneys, and
other vital organs
Source: unknown

Choline
Effect: Proper growth, fat
reduction; coloration
Deficiency symptoms:

Diseased kidneys and
enlarged liver
Source: Paprika

Calcium/phosphorus
Effect: Bone formation,
tooth formation
Deficiency symptoms: Poor
body development and
degenerations
Source: Fish and bone
meal, plant flour, peas,
spinach, carrots

Trace Elements
Effect: Bone and tooth
formation
Deficiency symptoms: Poor
development
Source: Fish and bone
meal, peas.

Handfeeding
frozen fish
food is the
most practical
and enjoyable
method of
feeding frozen
food to one's
fishes.

WHAT FOOD ORGANISMS FOR WHICH FISH?

Food Organism	Fish Size	Particularly suited for
Artemia nauplii	First food for all cichlids up to 4 cm	Eagerly taken by juveniles.
Cyclops	First food of mouthbrooder juveniles; for egg-laying cichlids only from 10th day onward to about 5 cm	As above.
Fish and crab roe	Juveniles 2 - 5 cm	Juveniles of predatory species,e.g., *Neolamprologus, Haplochromis, Julidochromis*, and others.
Bosmina	Cichlids 3-5 cm	All cichlids; often to 7 cm preferred.
Artemia	Fish size 4 to 15 cm	Treat for all cichlids.
Bloodworms	Cichlids 5 to 15 cm	Omnivorous and carnivorous species; to be given in small amounts only; eagerly taken.
Mosquito larvae	Fish size 5 to 12 cm	*Neolamprologus, Aulonocara,* small "*Haplochromis,*" *Julidochromis* and other small predators; eagerly taken.
Glassworms	Fish size 6 to 16 cm	All cichlids, especially plant feeders and grazers; eagerly taken.
Daphnia (Water fleas) 2 to 6 mm	Cichlids 4 to 12 cm (depending on size of food organisms)	All cichlids; variable acceptance, fish have to get used to food organism.

WHAT FOOD ORGANISMS FOR WHICH FISH?

Food Organism	Fish Size	Particularly suited for
Mysis	All cichlids 7 to 15 cm	Omnivorous and carnivorous fishes.
Krill (small)	Fish size 8 to 15 cm	As above.
Krill (large)	Fish size from 12 cm up	As above; or chopped up for small species;if chopped up,also for plant feeders (feed sparingly!)
Mussels	Cichlids from 15 cm up	Carnivores.
Shrimp	From 15 cm up	Aggressive predators.
Gammarus (freshwater shrimp)	Cichlids from 12 cm up	All cichlids.
Plankton	Juvenile fish (1-3 cm)	All cichlids (red plankton keenly taken).
Tubifex	Fish size 3 to 15 cm	Caution when feeding, since these worms may be contaminated by pollutants.
Earthworms	cichlids from 15 cm up	Omnivores and carnivores; or chopped up for smaller species variable acceptance.
White worms (*Enchytraeus*)	cichlids from 15 cm up	Omnivores (feed sparingly!!)

DRY FOODS

Modern aquarium management has made it possible for any aquarist to have a huge selection of foods and feeding opportunities at his or her disposal. Dry food has become particularly popular. It meets almost all nutritional requirements, so it can be used as the main dietary item. It is available in various types and particular sizes. The best known types are

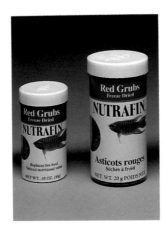

flakes, sticks (worm-shaped dry food particles), pellets, and tablets. Flakes are the most commonly used dry food, as most fishes show a distinct preference for them over sticks. The various flake sizes accommodate the food size requirements of most fishes. Since cichlids, just like other fishes, have specific dietary requirements, selecting the correct food is of paramount importance. Most fish food packages give a detailed ingredient analysis; if not, the respective manufacturer will upon request provide the necessary details. When choosing food for your fishes do not be influenced by fancy product type names. For example, sometimes dry food (irrespective of what type) is offered as vegetable food for those fishes that feed on plants and epiphytic growth. Yet a quick look at the ingredients list clearly shows that the product type does not meet the dietary requirements of those fishes. Sometimes these ingredient listings barely deviate from a standard food which should meet the needs of all fishes. Therefore, it can hardly be a special diet for a particular group of fishes. Of course, the same applies also to many other types of food.

As already mentioned dry food contains all required nutritional additives to assure full and complete functions of all organ systems. However, carotin, essential for proper coloration, is often absent from such foods. Natural carotin occurs only in live and frozen food organisms. During the heating and

Many different freeze-dried foods are available at pet shops. Those made primarily from whole animals, such as the Hagen red grubs shown here, are especially appealing to carnivorous species.

drying processes, however, carotin is destroyed. Therefore, some manufacturers add carotin again later as a chemical. When you purchase fish food make sure it actually contains the vital carotin. This substance contributes significantly to a full and complete development of proper coloration in fishes, a point that cannot be emphasized enough. Since some food organisms contain significant amounts of carotin, I will come back to this point later on.

There are two food characteristics that are important for cichlids; first, on the basis of the ingredient listing a food must be suitable for the species to be kept and the fish must eagerly accept the food. The price for such food is not necessarily the decisive factor when it comes to a qualitative good food or one that is unsatisfactory. It is therefore quite possible to be able to purchase an excellent food at reasonable

Altolamprologus calvus (shown here) and *A. compressiceps* like to use large marine shells as hiding places and as brood sites.

cost as well as having to pay a lot of money for an inferior product. But there are also other criteria to be observed when selecting a dry food. The food should be as fresh as possible (taste deteriorates and vitamins are destroyed during prolonged storage). Fishes are not keen to feed on old food. Moreover, dry food must not make the aquarium water cloudy.

It is common practice among fish food manufacturers to enhance the appearance of dry foods with food dyes. Different colors can mean variable ingredient composition or they can be purely esthetic. There do not seem to be any adverse effects on fishes from dye additives in food. I have even observed that fishes seem to prefer colored food over those that are plain, and may assume that the color of the flakes is a significant feeding stimulus.

FOOD ORGANISMS

Apart from dry foods there is also a wide selection of animals available as food for cichlids. These food organisms can be fed either alive, dried, or fresh-frozen. Most food organisms that are commercially available are traded in frozen form. Independent of how these organisms are to be used as fish food, I would like to briefly introduce the various organisms to you. Let us start out with the smallest of the food organisms commonly used for rearing small cichlids.

The first food for newly hatched fishes often is the nauplii (newly hatched young) of the saltwater crustacean *Artemia salina*, among the best possible food for such small fishes. Brine shrimp are rich in protein in the live as well as in the frozen stage, and they also contain a lot of carotin. The latter is very important for the proper development of coloration in fishes as they grow up. Brine shrimp nauplii are usually available in the freshly frozen state. Instead, you can easily raise these nauplii yourself with fairly simple and inexpensive equipment. In fact, it is even available in ready-to-use culture form. Good hatching rates are attainable with culture containers that are attached to the inside of the tank. They utilize the warmth and light from the aquarium, which facilitates optimal larvae hatching results.

You can also make your own brine shrimp hatchery! For that purpose you need an empty, clean 1-liter bottle, some air hosing, a small air stone, and a small vibrator pump. The bottle is half-filled with conditioned aquarium water. Then sea salt is added (sea salt is important for a good hatching result; please, do not use table eggs). Thereafter, the bottle should be placed in a bright location for the next few hours. The water temperature should be at least 26°C for the duration. Most larvae will have hatched after 24 hours and

salt or salt with iodine additive). For 500 ml of water you add 2 level teaspoons of salt. The air stone with air hose (connected to the vibrator pump) is placed on the bottom of the bottle and the salt water solution is vigorously aerated with large air bubbles. We now add the brine shrimp eggs (1 to 2 teaspoons full of can then be used as fish food. While this method of rearing brine shrimp larvae is very cost effective it is also somewhat awkward, because it is difficult to separate the larvae from still unhatched eggs inside the bottle. Since small cichlids must not damage their intestinal tract with unhatched *Artemia* eggs or the empty eggshells, it is

The photographer was considered to be an enemy, so the female immediately gathered up her young. The young will actually swim into the mouth of their mother ... and quickly disappear.

Obtaining a good hatch from brine shrimp eggs can be helped along by use of hatching devices and salt mixes.

advisable to purchase a professionally designed *Artemia* culturing device. It utilizes the warmth and light of the aquarium, since it can easily be secured inside any aquarium. It consists of three parts. A wide piece of plastic pipe is closed off at the top and at the bottom by two cylindrical plastic parts. Air is added from below to keep all *Artemia* eggs in constant motion, which assures an even hatching. Salt and *Artemia* eggs are added from the top. The nauplii can be removed after 24 hours without any

Adult brine shrimp. Living black worms. Living tubifex worms.

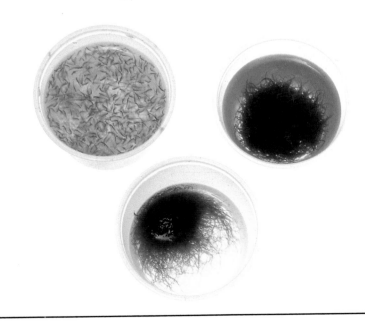

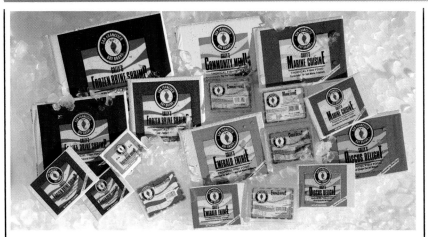

Frozen foods using a wide variety of food animals and in a number of different package sizes are available.

difficulties. To do that the air supply is turned off and the water is left standing for a few minutes. By then the heavier, unhatched eggs will have settled out at the bottom. Following that the *Artemia* nauplii gather in the tube and along the bottom of the hatching device. The dangerous shells of the hatched eggs will float to the surface. When attempting to remove the nauplii, the unhatched eggs are first siphoned off and discarded. This is followed by removing the clean larvae (without any shells) which are then immediately fed to the small cichlids. This method is simple and easy. Cleaning the device is similarly easy. All parts are quickly disassembled for a thorough cleaning. When using the traditional bottle method cleaning becomes very difficult, and since not all residue can be removed,

the subsequent hatching results get lower and lower.

Rearing *Artemia* nauplii all the way to the adult size of 10 to 12 mm is possible, but becomes quite involved. The nauplii are reared in an aquarium-size tank with aeration, heating, and a special food (commercially available). This method also requires sea salt. Since rearing such excellent fish food requires a lot of space, I can only advise against it. To satisfy the food requirements of a single aquarium would require an entire battery of brine shrimp rearing tanks. Yet, breeding and rearing *Artemia* can indeed be a hobby in its own right.

Because of the quantities required *Artemia* can really only be a supplementary food. Buying frozen *Artemia* is much simpler. Adult brine shrimp are eagerly taken by all cichlids. Even large predatory species will hotly pursue this relatively small food. Adult brine shrimp also have the same advantage as their larvae; they are rich in carotin, which is preserved when the brine shrimp are frozen. Since carotin contributes significantly to the development of full coloration in fishes, *Artemia* should be an essential element in the diet of your fishes. Now back to some of the other suitable food organisms.

Another excellent type of live food is the freshwater crustacean *Cyclops*, which are somewhat larger than *Artemia* nauplii. Although most newly hatched cichlids are large enough to start feeding on cyclops right away, there are a few exceptions. Yet, at an age of two weeks all newly hatched cichlids should be able to handle this food. Cyclops, too, contribute to the development of proper coloration in fishes and are an eagerly taken food.

Your fishes' menu can be further expanded with another crustacean, *Bosmina*. They are especially useful for cichlids in the size range from 2 to 6 cm. These extremely tasty miniature crustaceans will immediately be accepted by your fishes. Young cichlids from a size range from 3 to 4 cm upward can also be fed adult *Artemia salina*. Small predatory species can be reared with fish roe (1 to 2 mm diameter eggs). Lobster eggs (1 to 3 mm diameter) also contain valuable carotin. These orange-red eggs are suitable for feeding small fishes in general. Since they are also very protein-rich, both types of eggs should be fed sparingly only to grazing and herbivorous fishes in general.

Fishes 5 cm and upward can also be offered glassworms, bloodworms, and mosquito larvae as food. Mosquito larvae in particular are rich in vitamins. For sexually mature fishes they are considered to be a triggering stimulus for spawning. The highly nutritious larvae of phantom midges (*Corethra*, a type of mosquito) can, however, be dangerous to small fishes. These insects, which are essentially biting mosquitoes, have jaws developed as a stinging and sucking mechanism that can injure the digestive tract of juvenile fishes. This potential problem can be avoided by feeding frozen mosquito larvae. Many cichlids will eagerly accept mosquito larvae without much hesitation. Because of their excellent

A brooding *Tropheus duboisi* with a baby fleeing back into the mother's mouth.

nutritional value it is highly desirable to include them in your fishes' menu.

Lately, bloodworms as fish food have become a hotly debated item. Because of their taste (to fishes) and their bright red coloration, bloodworms are immediately accepted by virtually all fishes. On the other hand, they have the unfortunate characteristic of absorbing all substances dissolved in the water they live in. They reproduce prolifically in polluted waters and they are very adaptable. Since many natural waters are already more or less strongly polluted, bloodworms are not a recommendable food item for our fishes. Imported bloodworms, especially those from Asia,

Daphnia are an axcellent food for aquarium fishes.

are commercially bred in ponds. Although these larvae are not as attractively large and bright red as our local product, they may be less polluted. Bloodworms are not suitable as food for plant feeders or for grazers. On the other hand, for omnivorous and carnivorous species they are quite acceptable if they are occasionally included in the diet. The digestive tract of the latter cichlids is less sensitive to pollutants than that of generally herbivorous species. Using the capability the bloodworms have of absorbing substances, the aquarium trade now offers bloodworms that have been fortified with antibiotics. Using these larvae when treating diseases is, however, not advisable. First, it is uncertain in what concentration the medications have been taken up by the food organism. Moreover, the given doses also become equally uncertain. Treatment should therefore be initiated with other medications.

The third type of larvae that is a popular fish food is the so-called white mosquito larvae, commonly referred to as glassworms. These larvae occur only in extremely clean waters, so they are a very safe type of live fish food. Unfortunately, though, due to increasing pollution these larvae are harder and harder to collect. This makes them increasingly expensive to purchase. Glassworms are predators, which makes them a very nutritious fish food. When fed to sexually mature fishes, glassworms contribute significantly to the formation of eggs in females. They are often available as live food during the winter. They can easily be kept alive for days in a little water and stored in a refrigerator.

Another type of live food are water fleas (*Daphnia*). Just as with *Cyclops* and *Bosmina* the coloration of water fleas can be highly variable and is dependent upon their principal food source. Greenish *Daphnia* have fed principally on algae, but orange to reddish colored *Daphnia* are more nutritious, since

they feed mainly on the dissolved waste materials in ponds frequented by water fowl (ducks, swans). These miniature crustaceans also provide a lot of dietary bulkage and are an excellent supplementary food item for cichlids. Apart from the smaller *Cyclops* and *Bosmina, Daphnia* are available as different species, ranging in size from 2 to 6 mm. This variable size range makes it possible to feed *Daphnia* to a wide range of fishes. Therefore, it is better to use either small live food organisms or frozen food. For herbivorous cichlids it is especially important that the water fleas fed are not too large, since these cichlid species can not

Tropheus moorii, the Red Rainbow variant from Kasanga, Tanzania, Lake Tanganyika.

chew *Daphnia* or digest their strong chitinous shell.

Cichlids from about 10 cm onward can be given larger food organisms such as shrimp, clams, and bigger crustaceans. Small freshwater shrimp (*Mysis*) are highly nutritious. These food animals, which are about 8 to 12 mm long, are readily eaten by medium size to larger omnivores and carnivores. Herbivores (plant eaters and grazers) should only be fed *Mysis* very sparingly. Most of these fishes do not seem to be too keen on these little shrimp, anyway. Medium size to large marine shrimp can be used as food for large predatory cichlids. Some of these shrimp will even be taken by certain plant feeders and grazers. Clams are an excellent food for medium size to large predators; they can be given whole or chopped into smaller pieces. Since shrimp and clams are available in different species, you should try all of them on your fishes. You will quickly observe distinct feeding preferences among your fishes for some species over others. The same applies to krill, which vary in size from 6 mm to 30 mm.

All food organisms from the freshwater environment are particularly rich in

The older cichlids become the greater the size difference between the sexes. The males of *Cyphotilapia frontosa* will get cranial hump as adults (male above, female below). The depicted specimens originated from the same clutch.

minerals and trace elements. All freshwater crustaceans, such as *Cyclops, Daphnia,* and *Gammarus,* contain important carotins that are essential for proper color development in fishes. *Gammarus,* much larger than *Cyclops* and *Daphnia,* are also eagerly taken by fishes once they have become adapted to them. They are particularly useful as a carotin supplement when supplies of *Artemia* are temporarily unavailable.

One of the preceding tables presents a convenient overview of available food organisms. It is intended as a buying as well as a feeding guide for your East African cichlids.

VEGETABLES AND FOOD MIXTURES

As you can see there is a large food selection available to you. In fact, the food spectrum can be further enlarged. The dietary plan for herbivorous fishes can be enhanced with various types of lettuce, including green leaf lettuce and even spinach. Just prior to feeding the individual leaves are briefly immersed in boiling hot water. Then the leaves can be weighted down (planting lead, small rock, etc.) and placed on the bottom of the tank. It is important, however, that vegetable matter be as free as possible of harmful substances. Fertilizers or pesticides are particularly dangerous.

This feeding method also requires a bit of patience, until the fish has "sniffed out" the food and accepted it. For plant feeders and for grazers this is not only a valuable food supplement (bulkage!), but feeding on it also provides considerable activity for cichlids. Scalded lettuce leaves are actively taken by *Tropheus duboisi*.

In addition, various food mixtures can be prepared and offered, some of which are available from aquarium and tropical fish shops. Home recipes, however, provide for more diversified application since the aquarist has greater latitude in terms of composition (ingredients) and particle size. The following ingredients can be used: fishes, clams, prawns, shrimp (shells and all hard parts to be removed), scalded lettuce or spinach, boiled carrots and pasta (noodles), as well as peas (shelled). You can also add some egg yolk, red paprika (skin and seeds removed), calf and turkey liver (in small quantities), fish and bone meal, plankton, carotin meal, and, lastly, boiled (small-grain) oats. Freshly caught food organisms can also be added to this sort of food mixture. But defrosted food organisms should not be used, since some of the nutritional value has been lost.

All ingredients are macerated in a standard household blender. Frost-resistant vitamins can also be added. Finally, the food mixture is gelatinized (with white gelatin available from supermarkets, etc.) and portioned out into small freezer bags. In order to be able to break off small amounts of mixture later on, each freezer bag should contain only a small amount (when placed on its side a relatively thin layer should be formed throughout the bag). Each bag is first placed into the refrigerator to give the gelatin a chance to solidify and so bind the mixture together. This step is important so that the aquarium water later on does not become cloudy

during feeding time. After that the mixture is frozen and at -18°C it can be kept for up to 12 months.

The ingredients for each mixture can be selected precisely to meet the specific dietary requirements of a particular cichlid species. Preparing such mixtures is cost-effective although it is somewhat involved. All ingredient used must be fresh and the mixture thoroughly blended. Depending upon its composition, the ready mixture can be used for all cichlid species.

Meat such as beef heart or liver should only be used sparingly. Here we have to keep in mind that the digestion of warm-blooded animal meats (beef, pig) is somewhat inadequate in cichlids. Therefore, it is

A magnificent example of a juvenile *Tropheus duboisi* from Lake Tanganyika. Photo by H. J. Richter.

better to use fish fillets or other lean fish pieces (marine fishes). Apart from many minerals and trace elements, marine fish meat contains protein, which can be fully utilized by our cichlids. Liver used in the preparation of feed mixtures should preferably come from poultry and must be used in small quantities only. Finely chopped fish fillets can also be frozen. This is a type of food principally used as the main diet for predatory species.

FEEDING

From among the available dietary possibilities, live food organisms are clearly preferred. They must therefore complement the main diet as essential tasty tidbits. Feeding live food to your fishes carries with it the inherent danger of transmitting diseases. It is safer to feed frozen food organisms. The freezing process destroys a large number of the bacteria and after 72 hours all fish parasites are also killed. Anyone insisting on using live food should only use organisms out of ponds that do not contain any fishes. But even under these circumstances there can still be disease-causing organisms that can be transmitted. Ducks, swans, and other aquatic birds can carry disease pathogens from one body of water to another.

An important element in feeding is to distribute a daily ration over several feedings per day. As regards plant feeders and grazers, this method

Whatever live brine shrimp is not used within a day or so should be frozen while still fresh so that it can be safely used at a future time.

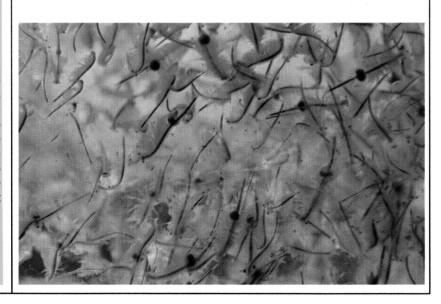

approximates their natural feeding regimen. For adult cichlids three feedings a day should be the minimum, but it is better to spread out the daily ration over 4 to 5 feedings. Juvenile fishes require food more frequently and should be given small portions 5 to 7 times a day. It is important that during each feeding every fish gets its share, but no food must be left over on the bottom. For the inexperienced aquarist it may be easier during the first few feedings to start out with very little and then add on small portions of food in order to get a feeling for the correct amount that should be fed.

Evenly distributed, adequate rations contribute to preventing aggression. I have observed frequently that cichlids—after having been on a reduced diet, especially during vacation time—became distinctly more aggressive. One reason for such aggression has already been described as food competition among the tank occupants. On the other hand, adequate feedings tend to reduce aggression since there is sufficient food for all cichlids. It is, however, important to make sure that your fishes are not being overfed. For those aquarists who hold down a

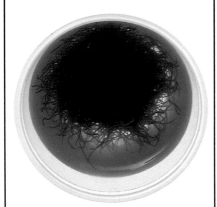

These black worms are an excellent source of protein for the more carnivorous Rift Lake cichlids.

regular job it is often difficult to feed their fishes regularly several times a day. For these people an automatic feeder provides the ideal solution. If someone leaves the house early in the morning, the tank light can be left on for 2 to 3 hours. A time clock will regulate the illumination periods. Aquarium fishes should be awake for at least 30 minutes prior to the first feeding. This makes it

A tubifex worm dispenser.

possible to offer the first feeding early in the day. Optimal fish nutrition is based on a variable diet including all essential nutritional components. Therefore, the various feedings should alternate with dry foods and live food organisms. Dry foods can be given in the morning and at noon, whereby the latter feeding can be postponed until the afternoon. Anyone getting home late can arrange to have the tank lights left on longer, and reduce the early morning illumination period so that the total illumination period does not exceed 12 to 14 hours. In the evening when you have time for your hobby you can offer the real choice tidbits in the form of live food organisms. Of course, there can also be a second feeding of live food organisms, e.g., of water fleas at mid-day. But the first feeding should always be with dry food, otherwise the fish will settle for the choice tidbits only.

Some of the predatory species, especially when wild-caught, will not accept dry food at all; but this is an exception. Clearly then these fishes should be given live food organisms instead of dry food.

It is advisable to give only one type of live food organism per feeding, otherwise the fish will select particular items and reject others. The most suitable food—from a nutritional point of view—is not always the most preferred; this holds true for humans as well as for aquarium fishes. "Variety" is the operative word in fish nutrition. Roughage and carbohydrates are just as essential as proteins.

Too much food leads to excessive fat deposition in and around vital organs. This is followed by metabolic disturbances. The organs affected, such as the liver, are the most important metabolic centers of a fish's body and can no longer properly digest all food taken in. Fishes with excessive fat deposits in these organs will lose all condition and become thin. Under such circumstances growth will nearly or completely cease.

A certain feeding regularity also brings other advantages to your fishes.

They will quickly adjust to any feeding schedule and will then wait impatiently for you at the front glass of the aquarium. Take advantage of the wide spectrum of fish foods available. Even the various dry foods can be mixed together into unique food blends. A variable diet increases appetite and lessens any dangers of nutritional deficiencies. In order to maintain peak coloration you should feed carotin-rich food organisms (*Cyclops*, *Daphnia*, *Artemia*, or *Gammarus* depending on the size of the fish) three to four times a week. Certain types of food that are not eagerly taken can often be made more palatable through gradual adjustment. Fishes will often accept certain foods only after they have tasted them frequently. This applies to all types of fish food. Unaccustomed types of food should only be given sparingly until the fish are used to them. This could be done just prior to the regular main feeding, when the fish are hungry and virtually ready to eat anything. If the fish continue to spit out the new food, the regular food should then be offered after a little while. After the new food has been spit out a few times the fish normally become accustomed to its taste and will then usually accept it.

With a correct diet cichlids will be highly active and feed vigorously. Make sure that all fishes get their share of the food. This also gives you an opportunity to check whether all fishes are healthy. Frozen food should be thawed out in a little sieve for better food dispersion, but it is also possible to feed frozen pieces provided the food is properly distributed. So far I have not been able to observe any adverse consequences from feeding frozen foods "on the rocks." Small food organisms will quickly thaw out in warm aquarium water, whereupon the fully thawed organisms will drop down from the floating (frozen) block. The advantage of this type of feeding is that hardly any food will get into the filter. Grazers often like to nibble on the thawing out food organisms as they come off the frozen block.

Sometimes the observant aquarist will notice that his fishes are clearly less interested in food. This is particularly noticeable among adult fishes. This is a manifestation of some slight overfeeding and can easily be corrected by reducing the daily food

Dry pellet foods are recommended as a supplement to your cichlids' diet.

portions for a few days (but NOT interrupting the overall feeding regimen). If this does not improve the fishes' behavior, one must assume the presence of a disease problem. Usually, however, the feeding behavior improves again after a few days of reduced food intake and the fishes will once again feed aggressively and eagerly. "Fasting" days can be instituted, although I am opposed to this; after all, there are no fasting days in the natural habitat. Fasting times occur periodically, anyway, since the aquarists cannot be present for all scheduled feedings. This is totally sufficient.

Feeding aquarium fishes during holidays and vacation times is nowadays no longer a problem, thanks to the availability of automatic feeders. These feeders can be adjusted in terms of portions and specific requirements of the fishes. Some automatic feeders are equipped with food dishes. In addition to flake food they can also be filled with food sticks, food tablets, and dried food organisms. Even medication can be administered in this way.

If vacation feedings are to be done by non-aquarists it is advisable to portion-out each feeding in advance. This avoids overfeeding and its undesirable consequences can be avoided. In general, however, experience has shown that an automatic feeder can feed the fish better than a non-aquarist. The latter will invariably give too much food. Therefore, it is better to use an automatic feeding device and have the neighbor look after the mechanical equipment.

BREEDING

SEXUAL DIFFERENCES

For breeding cichlids it is important to be able to distinguish between the sexes. This is the only way to pair up your fish. Unfortunately, though, for inexperienced aquarists the sexes of cichlids are relatively difficult to determine. This is easiest for species where the sexes have different coloration. Consequently, sexing adult specimens of the genera *"Haplochromis"*, *Aulonocara, Petrotilapia, Pseudotropheus, Labidochromis Labeotropheus* and *Melanochromis* can easily be done on the basis of external characteristics. While adult males of *"Haplochromis," Aulonocara,* and *Petrotilapia* are characterized by distinctive coloration, females are only insignificantly silver-gray. *Pseudotropheus, Labidochromis, Labeotropheus,* and *Melanochromis* also exhibit characteristic color differences between the sexes; however, here the females also show some conspicuous and attractive coloration.

Subadult specimens from these genera are difficult to sex. To determine the sex one can utilize the external features (egg-spots) as well as the size of the fish. Males nearly always grow faster than females of the same age. But this is, of course, only helpful when all juveniles are from the same clutch. Besides, many adult male cichlids have these distinctive egg spots. In fact, in subadults these spots may appear very early. Although females of some species also get egg spots, later in life these become less distinct than the ones in males. The most reliable characteristic for determining the sexes in cichlids is a comparison of the genital papillae, but this is, of course, only possible in subadults and adult specimens once the sex organs have been fully developed. Females can be recognized as such after they have spawned for the first time; thereafter the sexual opening remains distinctly enlarged.

One complicating factor in sexing cichlids lies in the fact that the genital papillae in cichlids can vary quite substantially. As an example, I would like to compare the *Tropheus*

moorii varieties. The so-
called northern and
southern varieties display
strongly variable genital
papillae. In males of the
northern variety, the
genital papilla—when
compared to the anal
opening—is not as small as
that of the southern
variety. The genital papilla
is usually as large as the
anal opening, but it can
also be smaller or larger.
On the other hand, families
from the northern variety
(Ikola, Bulu Point, Katonga,
and others) have a strongly
enlarged genital papilla,
which is usually three
times as large as the anal
opening. In these females
the opening is several times
larger than the anal
opening.

The above-mentioned
characteristics are also
found in *Tropheus polli* and
T. duboisi. On the other
hand, in the southern
variety of *T. moorii*
(Chimba, Chipimbi, all
rainbow variants, Moliro)
the genital papilla in
males is tiny when
compared to the anal
opening. Yet, females often
have an opening similar to
the anal opening, which
may also be slightly larger
than the anal opening. This
example shows that the
development of the genital
papillae varies not only
among cichlid species, but

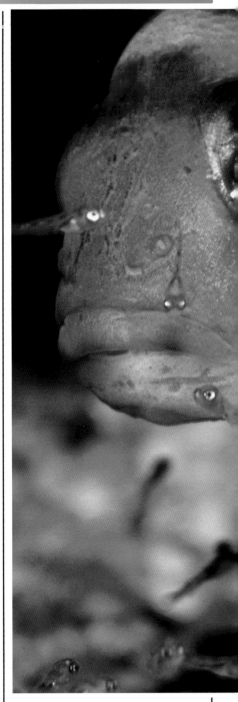

A female *Neolamprologus tretocephalus* guarding her newly hatched young.

also within subspecies and varieties. I have also observed that females that have not spawned for a while (for lack of a suitable partner) have an opening similar to that of males. Supposedly the female's genital papilla contracts again when there is no frequent spawning.

Being able to sex cichlids accurately requires much practice. Therefore, take advantage of any opportunity you have. When catching and transferring your fishes take a look at the openings of the sex organs. Take note of those animals where you know or suspect a particular sex. Then check when you have the fish in a net.

Unlike in Lake Malawi, where in many species the sex can be determined on the basis of external characteristics, this becomes the exception in Lake Tanganyika. Therefore, determining the sexes in juvenile fishes from Lake Tanganyika is virtually impossible. Suspected males often exhibit faster growth and greater aggression, and in some species an early change into adult coloration. For all these reasons it is advantageous when purchasing cichlids to get specimens of different sizes. This then gives rise to the possibility that the sex ratio balances out later on.

WILD-CAUGHT OR CAPTIVE-BRED SPECIMENS?

For anyone wanting to keep and breed cichlids, it is immaterial whether this is done with wild-caught stock or captive-bred stock. Usually, however, wild-caught fishes are more expensive than those bred in captivity. Moreover, freshly imported wild-caught fishes are sometimes more difficult in their care and maintenance. Often they are also carriers of diseases not found in captive-bred fishes, since the required intermediate host is missing in an aquarium situation. The acclimatization of these wild fish requires particular care and attention, and they usually prove to be rather susceptible to diseases. Therefore, beginning aquarists are advised against acquiring imported fishes until they have gained sufficient experience with breeding cichlids. Beyond that, there are no profound advantages in getting wild-caught fishes over captive-bred ones. On the other hand, wild-caught stock is essential when you need new blood for your breeding activities. But this aspect is really of interest only to the professional breeder.

With wild-caught specimens the coloration of individual animals can vary conditions than those occurring in the wild. Greater longevity, better growth (many fishes will get larger under aquarium conditions than in the wild), and more attractive colors are often the direct results from these

substantially. With modern aquarium management, combined with the enormous availability of technology, foods, and information, captive-bred fishes can indeed turn out far more attractive than those taken directly from their wild habitat. With these advantages you have a superb opportunity to provide optimum care for your fishes and in doing so produce even better technological developments. Of course, our breeding objective must always be to produce progeny that are compatible with the animals in the wild.

Generally, captive-bred fishes will breed easier than wild-caught stock. This is due to the fact that these fishes have already become adapted to conditions in captivity. Brood care is often

The male starts the courtship. The female touches the anal fin and genital papilla of the male with her mouth, where-upon sperm is given off.

particularly intensive and long-lasting in captive-bred females. This presupposes, of course, that a female is afforded the necessary opportunity and it does not exhibit any behavioral abnormalities (e.g., as can happen after repeated forced removal of immature larvae). Mouthbrooding females in particular often take advantage of aquarium conditions to practice extended brood care. Here the young are not swept away by currents and the aquarium is physically restricted so that the female can easily gather up all her young. Consequently, some females will take care of their young for several weeks. This sort of extended brood care is commonly seen in wild-caught animals. Once the young are fully developed they are released and never taken back into the mouth again.

In conclusion, I would like to point out that not every imported fish has just been removed from its natural habitat. Many cichlids are bred in Florida and along the shores of Lake Tanganyika. Since these fishes are bred in outdoor ponds they are often referred to as "lake-bred." Actually, these fishes should be considered as being captive-bred.

BREEDING MOUTHBROODING SPECIES

Many cichlid species are easy to breed since they will look after their eggs and young (= brood care) until the progeny can be released into their respective habitats. Only a few East African cichlids are reluctant to breed in captivity, and it is not the spawning act as such that poses any problems. Instead, the difficulties lie in establishing the correct conditions for a successful pairing. The principal problem lie in the well-developed aggressive instincts of many of the cichlids. It is therefore of paramount importance to have a properly interacting group of fish in a particular aquarium. Only when your fish feel secure and have a sense of well-being will they reproduce.

Placing the *right* individuals together in the same tank is the decisive factor for successful mating among cichlids. For that purpose mouthbrooders must always be in a larger group and substrate brooders in a smaller group. The latter are often monogamous and cannot be matched up at random with a particular sex partner. A certain selection within both sexes is of advantage so that permanent pair bonds can be established. Mouthbrooding species (maternal families without bonds to the father) can be easily bred in an aquarium provided the fish kept together are getting along with each other. Mouthbrooding females that are constantly harassed by some of the other tank inhabitants usually give up caring for their brood within a short period of time.

Courtship in all mouthbrooding cichlids proceeds according to a specific scheme. Adult males claim a territory in which they select a particular courtship site. The courtship site can be a patch of rocky substrate or a crater-shaped nest in the substrate (depending upon the species and its habitat in the wild). The courtship site is used as a base of operations for courting any approaching female. In fact, entire groups of females are approached in the hopes of enticing one of them to spawn. To persuade a female to mate the male "head butts" her flanks or even trembles

After a complete turn the female is now on the left and the male then touches her anal fin and genital papilla. The female releases the eggs which she picks up during her following turn.

Sexual differences: the genital papilla is located between anus and anal fin. In the male (shown here) the sexual opening can be clearly seen to be smaller than the anus.

In the female (shown here) the sexual opening is already distinctly larger than the anus. This female has not yet spawned. The genital papilla will get even larger after the first spawning.

Mating takes place inside a cave (opened up for photographic purposes). The female has already cleaned the spawning substrate and is now depositing her eggs. The clutch is then fertilized by the male.

before her. A female willing to spawn will follow the male back to the actual courtship site. There a more intense courtship commences with frequent head butting by the male.

When the female is ready the male will position himself on one side—in the middle of the courtship site—and remain there, his body trembling. The female approaches the male and touches his anal fin and genital papilla with her mouth. The male then turns away and the female positions herself to one side with body trembling. The male immediately touches the female's anal fin and genital papilla with his mouth. The pair then perform a few turns together (circling each other), which can best be described as "test" spawning. At that stage neither eggs nor sperm are given off. This procedure serves to synchronize both partners so that the eventual spawning can proceed without a hitch. The test spawning activities are usually frequently interrupted as the male breaks off to continue courting the female or to chase intruders away. Then the actual spawning

begins. Again, the male will position himself on one side with body trembling. The female follows with her mouth touching the male's side (at right angles). This is described as the T-position. During the subsequent turns both eggs and sperm are given off. Initially the female produces individual eggs and then more and more are extruded as the circling continues. Eggs are given off by the female as soon as the male touches the anal fin or genital papilla of the female with his mouth. The reverse takes place to initiate sperm ejaculation by the male. Immediately after the eggs have been given off, the female looks for them all over the courtship site and picks them up in her mouth. The sperm is also picked up by the female in her mouth during subsequent turns. After a while the egg supply of the female is exhausted. Although no further eggs are given off, the pair continues to circle and sperm is still produced by the male. This is still picked up by the female to assure that all eggs are fertilized.

Once spawning has been completed the female leaves

Once spawning has been completed the female starts to fan the eggs.

Once the larvae have emerged from the eggs they are then transferred to a new site by both parents.

the courtship site, but the male continues to court her. She may return a few times only to leave again. After a while the female looks for a secluded place in the aquarium and then remains there to incubate the eggs. During the first few days the eggs in her mouth must be constantly rearranged. Also, for the first 3 to 4 days the female will not feed. Females of those mouthbrooding cichlids that produce only a few eggs during each spawning start cautiously to feed again after a few days. Other cichlid females, such as those of *Aulonocara*, *"Haplochromis"*, *Petrotilapia*, *Xenotilapia*, *Callochromis*, and other species that produce large numbers of eggs or very small eggs, may fast throughout the entire incubation period. If a female has produced a large number of eggs and taken them into her mouth, some of the eggs may be eaten during the first few days. This only happens when the mouth is too full and the essential chewing motions cannot be made. Inexperienced females will often eat their first few clutches; the temptation to pick up food in spite of the eggs is simply too great. After several practice runs, however, even these females will "get the hang of it" and retain the eggs for the entire incubation period.

A female is seen here carrying a larva into the cave's interior.

During an incubation period of 3 to 4 weeks (depending upon the species), the eggs inside the female's mouth will have hatched and completely formed juvenile fish are carried. At that point the female looks for a quiet and protected place in order to release the young. If the female and her young are to remain in the breeding tank, tiny hiding places should be made available for the young. Yet, even in spite of such cautionary measures, large numbers of young will fall prey to other cichlids kept in the same tank. Yet, since most cichlid species will breed quite frequently, those few young that escape are often enough to maintain the aquarium population. Even in nature many young fish quickly fall prey to other fishes. If you want to make sure that there are enough survivors you should remove the required number of young from the breeding tank so that they can then be raised separately in another tank. The larger the young are when they are released by the female, the better their chances of survival, whereas the smaller the young the more likely they are of being mistaken for food by larger fishes. Those mouthbrooders that produce many small juveniles are particularly endangered.

There is also the possibility of transferring a

female and her brood to a separate tank just before the young are released. To do that the female must be caught in a net and then, without taking the net out of the water, gently maneuvered into a transport container. The female will sometimes spit out the young into the net or transport container. This, however, usually

Once all larvae have been transferred they are guarded by the female while the male defends the territory again.

the young into a separate rearing tank. Once the transfer has been completed, the female will usually take the young back into her mouth. A turned-over flowerpot or some rocky caves will give protection to the brood-caring female. If she feels threatened in any way she will not release the young. It is also advantageous to provide, apart from the cave for the female, additional smaller hiding places for the young (small caves, plant thickets, etc.). There are some mouth-brooding females which, after having released

happens only when the embryonic phase of the young has been completed and the female reacts in panic. Should this happen, the female must immediately be taken out, since the panic motions of such large fish can easily injure the young. The female can then either be returned to the main tank or be placed together with

their young, may then pursue them to prey on them. These include species of the genera *Callochromis* and *Ophthalmotilapia*, among others. In these species, the female and young should be separated after the young have been released from the female's mouth.

At this stage I must also

be honest and give those reasons that speak against transferring the female. If a cichlid species is kept in pairs only or in a very small group, is often difficult re-introducing such previously removed females. Once returned to her former tank the female rearing tank she often practices a long and intensive brood care. This can last up to two weeks and in some cases may even last for several weeks. The young will always be taken back into the mouth of the female as soon as she suspects imminent

is often considered to be an intruder by her tankmates. Therefore, such reintroductions must be done very cautiously and under constant observation. On the other hand, returning a female into a larger group should not present any problems at all.

When a female has been transferred to a separate danger. When brood care has ceased the female can be returned to her former tankmates. Mouthbrooding females that have been fasting during the incubation period can be brought back into prime condition by keeping them separate for a few extra days on a nutritious conditioning diet. It is an advantage to give a female

The larval stage is completed within a few days and the young fish start to swim about the tank.

the opportunity to recuperate from an extended brood care period. Not only does the female benefit from it, but also her progeny.

Since the incubation period among mouthbrooding cichlids varies widely, it is difficult to make any generalizations. Sometimes it is possible to extract relevant data about the incubation period in particular species from the literature (journals, magazines, books, etc.). To keep track of any breeding activities it is advisable to maintain detailed records or diaries, to be kept readily accessible.

Unfortunately, some aquarists remove the young prematurely and force them from the female's mouth, a practice that should be avoided. Forceful manipulation of the female's mouth parts can cause serious injury to the animal. Moreover, it has a serious debilitating effect on future brood care behavior. I have observed that mouthbrooding females that had premature embryos forcibly removed on several occasions eventually refused to carry eggs altogether. Years ago I received a group of cichlids from someone who habitually removed the eggs from mouthbrooding females about 10 days after the eggs had been laid. When I bred from that group the females always carried the eggs for 10 days only; that is, exactly up to that time when the eggs were originally removed. It took quite some time for the females to carry their eggs somewhat longer, and it was not until some years later that the females returned to their normal mouthbrooding incubation period. A number of the females, however, continued to eat their own eggs after about 10 days, a situation that never corrected itself in all those years. The effects of such methods on mouthbrooding cichlids has not yet been scientifically researched. However, the above example shows that brood care is not only genetically inherited, but also contains learned elements. Since not only the affected animals reacted adversely, the worst might be expected from successive generations. Natural brood care behavior must always be maintained and encouraged.

BREEDING SUBSTRATE SPAWNERS

Breeding any of the so-called substrate-spawning cichlids in a cichlid

As soon as the young of the *Neolamprologus tretocephalus* are free-swimming they need *Artemia* nauplii as first food. *Artemia* larvae are given directly into the school of young fish via the hose seen in the top of the picture. During courtship, spawning, and brooding periods the male displays very attractive coloration. All fins glow bright blue, as seen here.

A male and female *Pseudotropheus* sp. on the verge of spawning.

community tank is quite possible. But the free-swimming young are very much smaller than the young of mouthbrooding cichlids. Therefore, the progeny of substrate spawning cichlids is usually easy prey for other aquarium inhabitants.

Community tank-raised young cichlids have an excellent opportunity to form mated pairs among themselves. A bonded pair can best be recognized by the fact that the two fish tend to stay together, responding to each other's body signals, and they share the same territory without displaying any aggression toward each other. Other tankmates are driven off since they are considered to be intruders. When the pair is ready to spawn they engage in joint preparations for the upcoming courtship. The selected territory is fiercely defended—especially by the

male. Very large intruders are driven off by both partners. Protection through intensive defense of the territory is essential to prevent the eggs from being eaten by predators.

The female starts to look for a suitable spawning site within the territory. Then the spawning substrate is thoroughly cleaned by picking up all debris with the mouth. While this is going on the male continues to defend the territory. Depending upon the species involved, the spawning substrate upon which the eggs are going to be deposited can be rocks, slate, snail shells, marine clams, or simply the glass bottom of an aquarium.

Large marine shells with a suitable entrance for females are ideal for *Altolamprologus*. All snail and clam shells must be rinsed in hot water or even briefly boiled. Spawning activity starts after the spawning substrate has been cleaned. At first there is intensive body shaking and trembling. The eggs are eventually attached to the substrate by the female. Here, too, there are a few preliminary practice runs to make sure that everything is suitable.

As the male releases his sperm to fertilize the eggs, the female begins to pick up the eggs to begin incubating them in her mouth. Photo

A male *Neolampro-logus leleupi* guards its territory (above) from an intruder. It also guards it against conspecifics (below).

During or after the time the eggs are being laid by the female, the male fertilizes them. It is not necessary for the male to directly fertilize the eggs, but often the ejaculated sperm is fanned toward and over the eggs by the female.

Once the mating process is over, the female tends her clutch by constantly fanning fresh water over the eggs in order to avoid the development of fungus on the eggs. Eggs that are covered with fungus will be carefully removed and eaten. Depending upon the species the eggs develop in 1 to 2 weeks. Often the young larvae have to be "chewed" out of their eggshells. After the larvae have developed, some species will transfer them to a more protected site. There the wiggling larvae will remain and develop into juvenile fish.

The males of some species participate in brood care activities, but this is often refused by the female

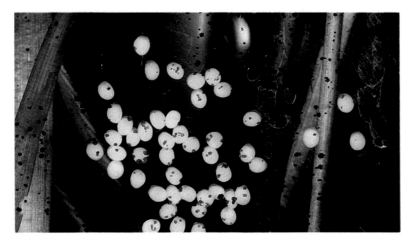

The male *Neolampro-logus leleupi* continuing to defend its territory (above). A close-up of the male (middle). A close-up of the eggs (below).

and the male is driven off. The territory continues to be patrolled by the male, while the female keeps the growing young under control.

With increasing growth the young get more and more adventurous, which makes the female's task more and more difficult. Finally, the female gives up and her progeny start to disperse. For a few days the young will remain in the proximity of the brood territory before they leave altogether. In order to protect the young against predators, they will have to be removed immediately after they begin to disperse. This can be done with a fine-meshed net or simply by siphoning them into a separate container. If the young are to be removed earlier than that (i.e. during the larval stage), some of them should be left with the parents, otherwise they may start to fight. A more or less intense pair bond always remains among monogamous cichlids. Often, however, the female is driven out of the territory after the maternal care has ceased. She will not return to the male's territory until the next courtship.

When breeding substrate spawners that deposit their eggs inside caves and other hiding places, it is important to remember that snails (e.g. *Melanoides tuberculata*, *Lymnaea stagnalis*, and others) can eat the eggs . Moreover, catfishes must be considered as nocturnal predators. Entire clutches of eggs can also be destroyed very rapidly by turbellarians. These are tiny white worms (the most common ones found in aquaria being planarians) that emerge from the bottom substrate at night and can often be seen after dark moving along the sides and front glass of the aquarium. Turbellarians can be eradicated quite naturally by introducing certain catfish species, such as *Pterygoplichthys gibbiceps* or *P. multiradiatus*, to the aquarium. It requires only a few individuals to rid the aquarium of them within a week to ten days. Caution has to be exercised, though, with large specimens of these catfishes, since they also feed on small fishes. The catfishes should be removed once treatment has been completed.

I would also like to point out that some cichlid broods require at least 26°C, better yet 27 to 28°C, during the developmental stages (and often beyond). Particularly sensitive in

A close-up of the female *Neolampro-logus leleupi* guarding her eggs.

this respect are lamprologine species (*Altolamprologus calvus* and *A. compressiceps*, *Neolamprologus tretocephalus* and *N. sexfasciatus* as well as others). Brood failure is frequently due to low temperatures.

REARING TANK

The thought of having to rear young fish usually does not arise until the first juvenile fish are already there. Ideally, a rearing tank should be set up early. For this we can use a tank from 80 to 100 cm long (possibly also larger or smaller, depending upon the species involved). This tank can be further subdivided with glass partitions into several large chambers. Such partitions can be secured by means of

Facing page: (Top) *Neolamprologus leleupi.* (Middle) A cluster of *Neolamprologus leleupi* eggs at various stages. (Bottom) The fry of *Neolamprologus leleupi.*

U-shaped aluminum strip grooves held in place with silicone. Glass sheets are then inserted into the groove. Depending upon space requirements individual chambers can be opened or closed. A heater placed in the center chamber is sufficient to provide enough heat to all chambers. A small filter is placed in each chamber.

CARE AND FEEDING

Filtration for a rearing tank must be of an appropriate size, otherwise the young may be sucked into the filter. In terms of decorations, etc., this tank should be kept relatively spartan. This is necessary to assure proper water circulation and to be able to detect quickly any excess food on the bottom. It must also be kept in mind that young fishes during their growth period (because of their elevated food requirements) need a lot of care, lots of water changes, and adequate filtration. Frequent water changes add new minerals and trace elements to the tank water that are vitally important for proper growth of young fishes. They are also particularly sensitive to poor water quality. However, should the nitrite level get too high and the water is of poor quality, do

NOT do any radical water changes. Some juvenile fishes are extremely sensitive to a transfer from nitrite-rich to nitrite-poor water (nitrite shock!). It therefore makes more sense to gradually return to the previous water quality level with several successive smaller partial water changes. During this period there should be little feeding or none at all in order to avoid a further deterioration of the water quality. *Altolamprologus calvus* and *A. compressiceps* require oxygen-rich, nitrite-free water for rearing their young.

The best rearing food for all cichlids is live *Artemia* nauplii. In addition, you can offer fine dry flakes, food tablets, and *Cyclops*. *Artemia* nauplii contain the essential carotin for proper color development; besides, they are also highly nutritious. The food given during the first few days determines growth, vitality, color, and ultimate adult size. We all like to hear compliments on how particularly healthy and attractive the cichlids are that we have raised!

Optimum care of the juveniles assures you of quality adult fish. Do not scrimp on food! Feed frequently and at regular

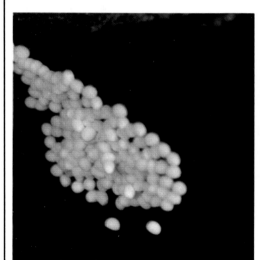

intervals (six to eight times a day), but small portions only so that there is no leftover food. An automatic feeder used for young fishes can be a valuable aid for aquarists with a 9 to 5 job. If such a device is not available you can make

young fishes are well cared for.

Since most cichlids will spawn every few weeks, it is simply not possible to rear all the young. You should rear only as many fishes as you (and others) may need in the future. Yet

A female *Oreochromis niloticus* caring for her free swimming young.

other arrangements. Live *Artemia* nauplii given early in the morning will stay alive for up to 6 hours (if they are not eaten early). Other substitute foods are food tablets that can be attached to one side of the tank. If put in place in the morning they will dispense fine food particles for a long period. This way your

it is a pleasant experience to observe a group of young fish grow from juveniles into adults, fish that would not have existed without you. In the final analysis, a successful rearing of fish always shows that you have provided optimum conditions.

A pair of *Aulonocara hueseri* from Likoma Island, Lake Malawi.

DISEASES

PREVENTION AND HYGIENE

Since most of the diseases occurring in aquariums are due to incorrect care and maintenance, let us first deal with the basic problems involved. Many diseases, especially those difficult-to-cure bacterial diseases, are caused by a lack of care and hygiene. The causes here are inadequate water changes, overcrowding, impacted aquarium substrate, as well as inadequate filtration. Immaculate cleanliness must be maintained in any aquarium through regular service and maintenance including all equipment items. Cleaning your aquarium must not become an ad hoc activity. Instead, this must be planned properly and several hours should be allocated for at least a single weekly cleaning, preferably always on the same day. Should this not be possible, an appropriate diary or calendar entry will remind you of the regularity of this task. This could take the following format: Friday afternoon=water change, filter cleaning to be done again on.....; major service at the beginning of each 3 month period, etc. This way

importatnt maintenance tasks are not forgotten. The quintessence of a successful aquarium is continuous and meticulous care in the maintenance program. This way many diseases are avoided right from the start!

Apart from regular partial water changes the bottom substrate must be thoroughly siphoned to remove all dirt particles. Gravel can be cleaned by means of a stand-pipe siphon, which actually rinses the surface layer. At the same time bits and pieces of floating or dead leaves are removed. Dead plant parts cause decay, which in turn uses up a lot of oxygen in the water. Filter media, such as synthetic wood, mats, activated charcoal, and bacterial substrates, must be rinsed out or exchanged on a regular basis. Wool and filter mats should be exchanged when the filter medium has turned light brown, an indication that waste products can no longer be removed or processed. In any event, brown-colored or even muddy filter media must be rigorously avoided. If the filter medium is clogged up, the aerobic bacteria can no longer process metabolic waste products into harmless substances. Water flow through the filter must be unimpeded and the filter must be able to "breathe."

If in doubt, it is better to replace filter wool early. Generally, filter media are not replaced simultaneously with a water change. It is however permissible to exchange part of the filter medium, so that most of the well broken-in filter remains unaffected in its function. Filter media, such as activated charcoal and bacteria substrates are regularly rinsed out with lukewarm water. Apart from regular weekly cleaning, the technical components should also be serviced and cleaned. The power head should be thoroughly overhauled every 2 to 3 months; this prevents a total, and usually unexpected, breakdown. At larger intervals the filter and filter vessel should also be cleaned in order to assure unimpeded water flow. While changing water, the glass sides of the tank should also be cleaned. All algae should be removed from the viewing side. Similarly, all cover glasses are wiped with a household sponge (with reinforced surface). All food remnants must be completely removed.

Should it happen that

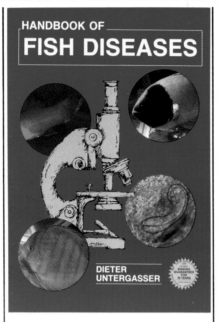

HANDBOOK OF
FISH DISEASES

DIETER UNTERGASSER

Handbook of Fish Diseases by D. Untergasser TFH TS-123, over 100 color photos. By any yardstick one of the finest and most useful books ever offered to aquarists.

during a particular feeding session you have overfed, all excess food must be siphoned off the bottom without delay.

A certain collection of cleaning implements, such as brushes, sponges, stand-pipes, siphon hoses, and similar items are essential items for aquarium maintenance and make this work much easier. But when working on an aquarium it is vitally important that all aids and implements are meticulously clean (kitchen clean!). All these items, as well as pieces of decoration, etc., should be used exclusively for that aquarium and be stored in a separate place. Buckets and other containers for aquarium water must never be used for carrying or holding harmful substances. Even when superficially clean, all containers must be thoroughly rinsed out before use.

TRANSPORTING FISHES

Buckets that have been used for carrying sand, gravel, and rocks can no longer be used for transporting fish. The inside of such buckets becomes roughened up so that it feels like sand paper; this could damage the skin of fishes. Instead, fishes should be transported in special fish transport bags. Large fishes must always be packed individually, i.e. one to a bag. Transport stress can cause the fishes to damage each other in transit. Beyond that, fishes should be kept dark and at a constant temperature. This can best be achieved with styrofoam boxes or even suitably padded (insulated) cardboard boxes. If the fishes have to spend several hours in transit, the bags should be filled with oxygen.

POISONING

Apart from the already mentioned causes for fish diseases in an aquarium, there can also be symptoms of poisoning. Toxic materials, such as lacquers, solvents, plastic foams, and similar substances, must never be stored in the same room where there is also a fish tank. If renovation work must be done involving these substances, the tank must be carefully protected (covered) against odors and vapors (all are toxic and many are water soluble!!). If need be, the tank can be isolated and sealed off (air tight) with plastic sheets. At the same time the air supply must be provided from an uncontaminated area (from an air pump in another room or on the balcony or terrace). Alternately, the renovation work area can be sealed off to avoid contaminating the aquarium. Wall paper glue, however, is non-toxic and can therefore be used without endangering the lives of your fishes.

Another type of air pollution that can be harmful to fishes is that caused by excessive cigarette smoke. Any aquarist who smokes must make sure that the room is well ventilated and excessive smoke must be avoided. Consequently, a fish tank does not belong in a bar or similar facility. With such pollutants in the air the aquarium takes over the function of a water pipe!...pollutants are absorbed by the aquarium water and clean air is given off again! These harmful substances remain in the water and will become concentrated until toxic levels are reached. Poisoning symptoms include distinctly elevated respiratory rates (rapid breathing). When your fish appear to have been poisoned, initiate an immediate water change or transfer the fish to unpolluted (but conditioned) water.

Another cause of fish poisoning can be dissolved heavy metals in the municipal water supply. Excessive levels of heavy metals leads to rapid breathing and liver damage. With continued exposure, the liver is one of the vital organs that is destroyed and the fish's abdomen swells up balloon-like within a short period of time (2 to 3 days). At that stage the fish is beyond help. Therefore, in areas where heavy metals in tap water are a problem it is strongly advisable to use a water conditioner in order to remove harmful metallic ions. Initially a short bath in water with water conditioner added may be indicated (dosage as per manufacturer's instructions). Since with poisoning of this kind fishes display symptoms similar to those observable for various bacterial diseases, a clear distinction is often difficult. In my experience, however, cases of fish poisoning seem to progress to a fatal stage much faster than with fishes affected by a bacterial disease.

Contaminated foods can also cause diseases. Dry food that has become damp can start to decay, releasing harmful decomposition substances. Dry food affected in this manner should be thrown away. All dry foods must be kept dry and stored in a sealed container. This also applies for live food organisms. Here we place a clean food dish next to the aquarium. Once the food is thawed out it must be fed immediately. Do not re-freeze excess food. The feed dish should be thoroughly washed after it has been used. Live food that is already giving off rotten odors is no longer usable.

If you keep your aquarium in meticulous condition and perform all required maintenance at regular intervals, then you will have already taken a significant step toward the well-being of your fishes.

BACTERIAL INFECTIONS

In the following paragraphs I would like to discuss the most common diseases of East African cichlids. The most frequently occurring diseases are those caused by bacteria. Depending on the type of disease involved, its manifestations can be fin and tail rot, mouth fungus, white, yellow, and red skin spots and patches, abnormal swimming behavior (shimmying, tail fin folded), rapid breathing (gills affected by pathogen), and/or a bloated abdomen. In cichlids the clinical picture is most commonly characterized by rapid breathing in conjunction with a bloated abdomen. It is not uncommon to find the bacterial infection to be secondary in nature, as a consequence of intestinal parasites (please refer to the section on intestinal parasites). Bacterial diseases can be fatal rather quickly, therefore early recognition is of paramount importance. The sick fish refuses all food and keeps apart from the rest of the group. Since any bacterial infection can be caused by a multitude of different bacteria the use of a broad spectrum antibiotic is recommended. The effective range is relatively wide and in many cases it works very well. The commercially traded medication nitrofuran is, because of its wide ranging effect, particularly recommended. This medication can be administered as a dip or bath as well as a long-term (chronic) treatment.

Bacterial diseases can also be treated with oxytetracycline HCl (Terramycin), which is present in some commercially available remedies. The dosage should be 4 g per 100 liters of aquarium water and the treatment should be maintained for three days. Upon completion of the treatment most of the tank water must be replaced and/or filtered over fresh activated charcoal. This latter step is to remove all remnants of the medication. This charcoal should be exchanged/replaced after 3 to 4 days.

Although oxytetracycline is also a highly effective medication, it tends to make the water fairly cloudy. During the first day of treatment the water is generally intensely yellow, on the second day it is yellowish-brown, and during the third day it increases in intensity to a reddish-brown. At that point the fish are only barely visible in the tank. This disadvantage, however, does not affect the otherwise excellent efficacy of the medication. With this medication we must also give a massive water change and use activated charcoal once the treatment has been completed.

Fish tuberculosis is another bacterial infection that tends to afflict East

African cichlids. It is caused by a species of *Mycobacterium*, and the disease is generally considered to be incurable. The symptoms are tuberculous inflammations of the skin, the underlying muscle tissue, and the internal organs. Further manifestations are a dark coloration, irregular swimming movements, and a distinct loss of weight. Research at various institutions has shown that a large segment of our tropical fishes are carrying this pathogen (without causing a disease outbreak). Fish tuberculosis can indeed be avoided, principally by providing optimal conditions. It has been shown that tuberculosis breaks out among those fishes in an aquarium that are not feeling well. This affects particularly those that are strongly suppressed by other fishes in the same tank. Even fishes kept on their own (without other specimens of the same species present) can also be candidates for this disease. It is usually difficult for the aquarists to recognize fish tuberculosis. Frequently, individual specimens just drop dead without any external signs. Infected fishes can also carry the disease for quite a long time before they eventually die.

Only after there has been a number of unexplained deaths does suspicion tend to be aroused. Fish tuberculosis is a transmittable disease. Infected fishes as well as all equipment (nets, filters, etc.) must not come into contact with healthy fishes. This bacterium cannot be destroyed by heat, cold, or an acid. It is an extremely resistant pathogen. Fish tuberculosis can also be transmitted to humans, where it manifests itself through the occurrence of skin lesions. The only fish known for its resistance to fish tuberculosis is the Blue Antennae Catfish (*Ancistrus dolichopterus*). This fish can transmit the disease, but itself appears to be immune to it. Therefore, it must not be transferred from a diseased tank to a healthy tank. In any event, appropriate prevention is always the best possibility to protect your fishes against this diseases.

GASTRO-INTESTINAL PARASITES

The second most common problem among East African cichlids involves intestinal parasites. This problem develops in an aquarium when its inhabitants are exposed to unfavorable conditions. Wild-caught specimens are particularly susceptible to this, since the long transit time has weakened their resistance. The symptoms of this disease are whitish-slimy to transparent feces, frequently together with coughing and regurgitating mouth movements. An affected fish becomes lethargic but will still accept choice tidbits. Later on all food is refused. The most effective treatment is the administration of the medication METRONIDAZOLE (available from drug stores and chemists). The aquarium is dosed with 400 mg per 100 liters of water for a period of three days. This is followed by a large water change and a subsequent use of activated charcoal (for 3 to 4 days). After that period the charcoal should be replaced. After the treatment the fish may not eat very well (or only reluctantly) for a few days. If the treatment is not successful it can be repeated.

Because of its effectiveness, Metronidazole is the preferred medication against intestinal parasites. Fishes affected by intestinal parasites often suffer from secondary bacterial infections. The bacteria can invade the body and attack healthy organs through the intestinal walls damaged by the parasites. If both diseases occur together it is advisable to first give a short bath with nitrofuran (to kill the bacteria), followed by a prolonged treatment with Metronidazole against intestinal parasites. These medications should not be used simultaneously.

FUNGUS INFECTIONS

Less frequent but often rather tenacious are fungal infections on fishes. These kinds of diseases can often be traced back to inadequate conditions. Fungal infections are often of secondary

A *Symphysodon aequifasciata* suffering from chilodonella.

nature, infecting sites on the fish's body previously attacked by other pathogens. They also occur when there has been mechanical damage to the mucous membrane. This can happen when fishes are being caught with a net, fighting (aggression!), or during transport. Fungal infections are easily recognizable as a whitish gray layer over the affected areas. There are a number of commercial products available from aquarium and tropical fish shops. Dosing instructions provided by the manufacturer for each product should be closely followed. A medication called 'Furamor' is particularly effective (available from aquarium shops). Alternatively, you can use a dilute solution of Mercurochrome (drug stores). To treat an affected fish it is caught (twice daily) in a hand net and removed from the tank. Then a cotton ball drenched in the Mercurochrome solution is gently dabbed onto the affected areas of the fish. As already mentioned, fungal diseases are often of secondary nature. They occur principally in conjunction with bacterial and parasitic diseases. It is important to treat both diseases simultaneously. A long term (chronic) bath is given to treat the bacterial infection while the fungus patches are treated (with Mercurochrome) twice a day by taking the fish temporarily out of the bath. The same procedures are applied when there is a concurrent appearance of a parasitic and a fungal disease. The combination of short and long baths together with the Mercurochrome treatment facilitates the simultaneous treatment of several diseases. Some species of East African cichlids are particularly susceptible to various fungal infections. The cichlids involved are mainly *Aulonocara*, '*Haplochromis*', sand-dwellers, pelagic species, and some *Lamprologus*. The eyes are particularly sensitive and susceptible. These fishes should only be handled with a very fine-meshed net, and should then immediately be removed from the net again and transferred (quickly) by hand or in a plastic transport bag.

GILL FLUKES

The invertebrate phylum Platyhelminthes (flatworms) has a number of species that are fish parasites. This phylum is subdivided into three classes: Turbellaria, Trematoda (flukes) and Cestoda (tapeworms). Of these only the latter contain some important fish parasites. Aquarists are principally concerned with trematodes (flukes) living parasitically on aquarium fishes. Flukes, which can attach themselves by special mechanisms to virtually any part of a fish, are often found in the mouth and gill cavity (=gill flukes) of fishes, including East African cichlids.

Flukes are not overly common on captive-bred aquarium fishes but can occur in considerable numbers on wild-caught and newly imported fishes. These parasites do not seem to transmit readily to the captive-bred progeny of wild stocks. Symptoms of a gill fluke infestation are gill covers wide open, rapid breathing, gasping for air, and rubbing the gill areas on objects in the tank ('scratching').

There are a number of medications available for the eradication of gill flukes, including a substance called trichlorfon (from drug stores, chemists). The most effective treatment is a bath for 1½ days in a concentration of 1 mg per 1 liter of water. This must be followed by a major water change and the use of activated charcoal, as already discussed earlier. Trichlorfon is toxic to fishes as well as to humans.

Therefore, it is important that this substance (in concentrated form) does not come into contact with your skin, eyes, and mouth. Another effective eradication agent is flubendazol. This should be given at a concentration of 200 mg per 100 liters of water. The treatment should last 3 days and is to be followed by a 50% water change and the use of charcoal. For the latest findings about this medication your attention is drawn to the book *Discus Health* by Dieter Untergasser.

Great care needs to be taken when medications are being used. Duration of treatment and recommended dosage levels must be maintained in order to avoid organ damage to the fish. You must also avoid as far as possible any direct (skin) contact with these substances. All medications on hand must be kept away from children and under lock and key. Unlabeled medication packages must be properly identified with content and purchase date. Commercially traded products have an expiration date (for maximum effectiveness of the active ingredient(s))and should be disposed of once this date has been reached. All medication should be stored in a cool (refrigerated), dry, dark place.

DISCUS HEALTH
SELECTION, CARE, DIET, DISEASES & TREATMENTS
FOR DISCUS, ANGELFISH AND OTHER CICHLIDS

DIETER UNTERGASSER

Medications should only be used when absolutely necessary. Some medications tend to lose their effectiveness once they have been used repeatedly in the same tank, because some pathogens can develop a resistance to the medication's active ingredient.

SALT BATHS AND ELEVATED TEMPERATURE TREATMENTS

Apart from the above-mentioned treatment methods there are still other possibilities to treat particular fish diseases. Health problems that can not be clearly diagnosed will often respond favorably to a salt bath (preferably sea salt) in conjunction with a temperature increase in temperature to 33°C. However, the genera *Julidochromis, Telmatochromis* and *Chalinochromis* are somewhat sensitive to such high temperatures. For these species 30°C should not be exceeded during the treatment period. The effectiveness of many medications will be enhanced with an increase in water temperature. Keep in mind though that any increase in temperature lowers the saturation level for dissolved oxygen in the water. Therefore, when the temperature is increased make sure that there is also adequate aeration (possibly add another air stone, if need be).

The use of sea salt is recommended for treating the initial stages of such diseases as fungus, white spot, and skin and gill flukes. Furthermore, a salt treatment can also be given at larger intervals for purely prophylactic purposes. To set up a prolonged salt bath use 1 heaping tablespoon of salt for every 10 liters of tank water. This treatment is maintained for three days in conjunction with a simultaneous temperature increase. A so-called short salt bath consists of adding 10 g of sea salt per liter of tank water. After a fish has been given a short salt bath,

For detailed information about the health of cichlids in general and discus in particular, the book *Discus Health* is highly recommended.

Hobbyists should familiarize themselves with the various kinds of medications available so that they'll know how to properly treat a sick or injured fish.

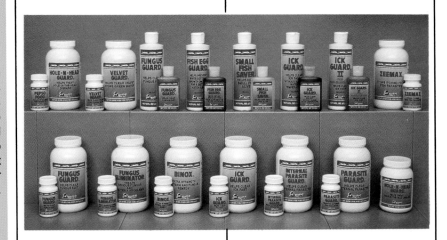

it should be gradually acclimated back to regular aquarium water.

Following a particular treatment, recovery can be enhanced by adding various preparations to the aquarium water. Water conditioners seem to have a beneficial effect following a fungal infection. These substance tend to expedite the recovery of the mucous skin layer. Various vitamin preparations given immediately after disease treatments can also speed up a fish's recovery. Since most fishes will refuse to feed while they are sick, feeding must be resumed cautiously and sparingly following any treatment. It may take days, sometimes even weeks, until fish resume their regular feeding pattern and food quantity. Appetite often improves when choice tidbits are offered.

It is extremely important to keep all implements used for treating sick fishes totally separate. As soon as a tank shows any disease symptoms all nets, filters, plants, and decoration as well as all cleaning equipment,is to be carefully kept away from healthy fishes and their tanks. Only after the fishes are healthy again can all these items and implements—AFTER they have been thoroughly disinfected (rinsed in hot water and permitted to air dry)—be used for healthy fishes and aquariums again.

INDEX